Working at Woodworking

This baking center is a good example of the type of cabinets Tolpin builds in his two-car-garage-size shop. It represents a fusion of American 'country' styling and European construction technologies.

Working at Woodworking

How to organize your shop and your business

Jim Tolpin

Cover photo by Patrick Cudahy

First printing: December 1990
Second printing: March 1992
Printed in the United States of America

A Fine Woodworking Book

Fine Woodworking® is a trademark of The Taunton Press, Inc., registered in the U.S. Patent and Trademark Office.

The Taunton Press, 63 South Main Street, Box 5506,
Newtown, CT 06740-5506

Library of Congress Cataloging-in-Publication Data

Tolpin, Jim, 1947-
Working at woodworking: how to organize your shop and your business / Jim Tolpin.
p. cm.
"A Fine woodworking book"– T.p. verso.
Includes index.
ISBN 0-942391-67-5 $21.95
1. Cabinet-work. 2. Small business– Management.
I. Title.
TT197.T63 1990 90-49037
684.1'6'068– dc20 CIP

I dedicate this book to Alicia Bates, riding life's other merry-go-round as we reach for the golden ring.

Acknowledgments

I wish to acknowledge the following good people, who, sometimes unwittingly, taught me something I needed to know: Sam Rosoff, Irving Pregozen, Jay Leech, Ludwig Furtner, Warren Wilson, John Muxie, Steven Souder, Johnny Coyne, Frank Whittemore, David Sawyer, Doug Fowle, Dan Valenza, Bud McIntosh, Diane Mayers, Ken Kellman, Gary Jonland, Daniel Neville, John Ewald, Chris Marrs, Charles Landau, John Maurer, and Walter and Susan Melton. If not for them, I'd still be holding the dumb end of the tape.

I am especially indebted to Francis "Nat" Natali, who showed me that learning to write was not unlike learning to ride a bicycle: "...just keep at it and you'll find that the balance comes of its own." And to my editor, Laura Tringali, who kicked off the training wheels. Finally, I wish to thank photographer Pat Cudahy, for his perseverance and excellent eye.

Contents

Introduction

It's been nearly 20 years since I first started working at woodworking. In the beginning, I worked alone out of a one-car garage outfitted with a minimal number of tools. Today I am the sole proprietor of a one-man cabinet shop, working out of a two-car garage outfitted with a minimal number of tools.

"That boy has gone far," I can hear you saying, "now he's got room for a second car." There is, however, an even greater difference: This boy can now afford a second car. In my first 10 years of working wood for a living, I could barely provide for myself, let alone a second car, or the family of four I now support (another difference 20 years can make). It was not the woodworking itself, but rather the way in which I was working the wood that forestalled my financial success.

During those first 10 years, I could often be found building highly refined pieces of casework that required much hand joinery and tedious fine detailing. I called the results of these efforts kitchen cabinets. With each delivery, I bathed in my clientele's enthusiastic approval of my work. Little did I realize how much of the celebration was due to their joy at obtaining such a piece at such a price—at least not until the day I looked up from my work and realized that this was not cabinetmaking. I was going broke, like every custom furniture maker I had ever met. It had come time to unplug myself from the shop for a while and try to figure out what I was doing wrong.

It didn't take very long before I had three likely answers. First, I did not have the faintest idea about how to go about building cabinets—in fact, I didn't really know what a kitchen cabinet was. Second, my methods and tooling, such as they were, were primitive and counterproductive. And third, I was an abysmally poor businessman. To continue to practice woodworking as a livelihood, I would have to understand the market's perspective of a quality piece of cabinetry and learn how to build it as efficiently as possible. I would also have to learn how to present myself and my products to the public in a successful way.

I began my education by taking a long, hard look at top-of-the-line cabinetwork on display at several local showrooms. I quickly found that the products imported from Europe had a great deal to teach me. With childlike awe, I examined the incredibly sophisticated hardware system that allowed these cabinets to achieve fitting tolerances generally unheard of in kitchen cabinetry. Looking closer, I learned that the casework was manufactured so that any variety of hardware elements could be installed: The same cabinet module could therefore hold numerous configurations of doors, drawers and shelving without further processing. My mind boggled at the implications this could have on my primitive approach to cabinetmaking.

Imbued with a new vision of the cabinetwork I would strive to produce, I turned my attention back to my shop. It would be necessary to revamp the layout of the workspace completely and upgrade the tooling. Out went dust-collecting workbenches and an ancient, massive surface planer. In their place I substituted mobile caddies for tools and materials, and a multipurpose knockdown work platform. A new, lightweight planer was hung from the ceiling, instantly ready for use with a heave on a rope. I regrouped the major stationary tools into symbiotic relationships, creating compact work stations; I also upgraded the table saw, radial-arm saw and drill press with new fence systems, which practically eliminated the need for a tape measure during most sizing and milling operations.

With the physical plant in shape, I then focused my attention on the production process itself. I analyzed the way in which materials flowed through the shop and how processes could be grouped and sequenced. I was mapping out a plan that would make the most of my time and my modest floor space. The result was the creation of a production flow chart that carried me smoothly from the initial stages of developing the layout and cut lists, through the production process, to the final installation of the product on site.

At this point, with a firm grasp of the products I intended to offer the market and a revitalized shop and work style, I needed to jump just one remaining hurdle: into the world of business. But the creation of a viable business enterprise struck me as something intangible and daunting. I soon realized, however, that I need only apply the same frame of mind that had wrought such miraculous changes in my shop.

I began by talking with people who were successfully operating small businesses like mine, and quickly learned what I needed to do. I explored the paper trail that must be followed to lend a bureaucrat's vision of credibility to a small business. I researched the legal ramifications of the various forms a business might take. I became acutely aware of the art of image-making. In dealing with clients, I learned not only how to clean up my act, I made one up. Finally, I designed a paperwork system specifically for small-shop cabinetmaking that simplified the day-to-day documentation of production data and business transactions.

This book is divided into the three major areas in which I worked to fix what was going wrong with my career: the shop, the process and the business. Together they comprise the story of how I now go about working at woodworking. It's the story of working hard and working well, and of producing a product whose market value amply rewards its maker. I am thankful to have been able to write this book; who knows, if I hadn't looked up from my work that fateful day, things might have turned out very differently. Not having learned what it has taken to get me to this point, I probably would have ended up going back to college for an advanced degree and would now be stuck with a "real" job. I shudder to think....

—Jim Tolpin
December, 1990

Section I

The main section of Tolpin's current shop is 22 ft. by 24 ft. on the inside, with a shed extension housing an office and dead storage on the south wall.

The Shop

There is probably nothing more subjective in the mind of cabinetmakers than the idea of what constitutes a perfect shop. But let me say this: Your perfect shop will be the one that works with you in your daily endeavors to create a product in an efficient, enjoyable manner. If your shop doesn't do this, it's worth your time and energy to see that it does.

The shop I currently work in is not large by many cabinetmakers' standards; it's basically a two-car garage with the blessing of a 10-ft. high ceiling. Over the years, however, I've learned that size is not the single determining factor in creating a workable environment. Other factors to consider (and believe me, we will) are efficient use of floor space, orientation of the major stationary tools and work stations, careful management of material flow through the shop and proper sequencing of production processes.

Over the past two decades, two very basic rules have evolved as a result of life in small workshops. The first rule is this: "Nothing gets to occupy the shop's floor space except the essential stationary power tools." Bikes, washing machines and camping gear are no longer tolerated in my workspace. Tools, shop accessories and storage systems are hung on the wall where possible; permanently fixed workbenches, nonfolding sawhorses and certain immobile power tools are taboo. The end result is an open, easy-to-clean workshop. Ample room exists at each work station, raw materials and components flow easily through the space and everything needed to produce the product lies close at hand.

The second rule that bubbled up out of the primordial sawdust concerned the production process itself. Simply stated, it is this: "No project will be assembled until there is a place outside the shop for it to go." Because assembled components uncannily resemble empty boxes, storing them renders vast amounts of space essentially useless. To minimize this small-shop gridlock, the production method I have developed over the years places assembly at the very end of the process. Allegiance to this rule simply requires that a project be scheduled for installation immediately after assembly. But this doesn't mean that only one project can be processed at a time. In fact, the block production method detailed later in this book actually makes it easy to do several jobs at one time.

With these two rules now etched in ebony, it's time to create your perfect shop. This section will cover location, utilities and structural considerations; layout of the major tools and work stations; and the tooling that you will need to make it in this business.

Chapter 1: Location and Workspace

Let's assume we are going to create a perfect shop from scratch. This is the best way to start, especially if you are currently stuck in a compromising shop situation. Besides, after you have read this chapter, you may decide you want to move to a new shop anyway.

Where to Locate

The first issue to consider is this: Do you want your shop to be at home or a commute away? Part of the answer will arise from an analysis of your personal habits. Some people can work effectively at home on a daily basis, whereas others have difficulty putting in a single productive eight-hour day. The problem is, of course, that the attractions of home life beckon just outside the shop's door.

It is, however, well worth the effort to learn to work productively at home. Overhead will very likely be much lower than for a separate shop, and there will be no commutation expense. Even more important is the coherence that can come into your life when home and work form a symbiotic relationship. It's part of the old American dream, and part of the gift of being a cabinetmaker that is worth experiencing.

Unfortunately, it may not always be possible to have a shop at home. Consider the following before deciding:

Unless your name and reputation are already well established, your shop should be close to the region's job market. If your home is an hour's drive over poor roads from the nearest town, your clients really have to want to give you their business to justify their passing all those other cabinetmakers on their way up to your place. Of course, the better your work is and the more competitively you price it, the farther they will be willing to drive, but only to a point. And if your home is not on the client's side of that point, you may not be able to have your shop there.

The trucks that will come and go from your shop's door will need good physical access. The driveway must neither swallow vehicles and traveling salesmen every time the rain turns it to mud, nor be so steep that the strain of climbing it turns trucks' driveshafts into wet spaghetti. There must also be a place for vehicles to turn around.

Another important consideration is whether the law will allow you to work in your home. Contact the city or your regional planning office to find out whether your property is zoned for commercial use. Or simply go into the town hall and ask to see the zoning map and the list of attendant ordinances. (Also check to see if there are noise restrictions that might affect you.) As a one- or two-person shop, you will be small, but you won't be invisible. If there's a legal problem with your presence, you might eventually find yourself out of a shop.

One final question: How much can you afford to spend for shop space? Unless you are lucky, be prepared to pay good money for a good shop location. If your shop is at home, you have a distinct financial advantage, but if you find it difficult to work productively there or

The cabinet shop of John Maurer, of Carlotta, Calif., is just a little bigger than a single-car garage. It's about as small as you can go and still maintain sanity.

if access or zoning are potential problems, then this advantage may turn out to be illusory.

The rent or mortgage on a shop space is a fixed burden that will affect your ability to turn a profit. But locating yourself in an area that affords good visibility to an appreciative clientele will contribute greatly to your ability to carry a higher overhead without jeopardizing your profit margin. If you're just getting started, you'll probably have to accept a lower profit margin to be able to afford a good location, but make sure you're at least covering overhead (see pp. 118-119) and earning a wage comparable to that earned by cabinetmakers at a local commercial shop. Try not to compromise on the shop if you can help it, or you may find yourself slowly sinking into the sawdust over a number of years.

Size and Space

If it's legal to work there, if the driveway doesn't eat trucks and if your clients don't ask "What state are you in?" when you give them directions, then you may have a good location for your shop. It's time, then, to take a close look at the physical plant, considering such factors as overall size and floor configurations, placement of doors and windows, availability of utilities and possible interior structural modifications.

A good shop is not necessarily a large shop. In fact, I rate ease of access and high visibility far ahead of voluminous floor space. I've found that a shop as small as 528 sq. ft. (a typical two-car garage, as shown in the photo that opens Section I of this book) is entirely adequate for a one-person shop. Too much space can be detrimental. A shop approaching 1,350 sq. ft. puts extra mileage on your boots as you trudge from one work station to the next. In addition, excess materials all too easily find their way into a large shop and impede production flow; and you find yourself heating, lighting and paying for more space than you really need.

This is the massive shop of Wallace and Hinz, of Arcata, Calif., builders of fine custom bars.

The configuration of the space often determines how well the shop will work for cabinetmaking. A square or rectangular space is ideal– L-shaped or joined spaces with access through partitions tend to work against efficient production. In my experience, an open square can be used far more efficiently than an unusually configured space of twice the square footage.

In addition to the open shop area, it's nice to have a bathroom and perhaps a bit of covered dead storage. A separate office is a necessity. This quiet, dust-free space is the place to make telephone calls; talk to clients; do design work and bookkeeping; store and use books, portfolios and product samples; have lunch; and change out of your workclothes. If an addition to the shop is out of the question, consider building an enclosed loft in a high-ceilinged shop, or sacrifice a 6-ft. by 6-ft. space in the corner of the work area for an office.

Lighting and Electrical Requirements

It's worth noting whether the structure you're considering receives ample natural light. It's probably not sufficient reason to turn down an otherwise good space, but a shop buried deep in the woods or surrounded by tall buildings can be gloomy to work in. Ideally, the shop will be exposed to open sky in at least one direction, and the windows will be situated to take advantage of it. They should be located as high as possible on the walls to keep direct sunlight from playing across the work stations. (The best condition is to have most of the windows high on the north side of the building, and a good number of opaque skylights in the roof.)

Unless your shop has a waterwheel and jackshafts, or you're just doing this for fun and aerobic conditioning, you are going to need electricity. (See the wiring diagram for my shop on the facing page.) Ideally, look for at least 100-amp service on its own meter to feed separate lighting and socket circuits for the major stationary tools– you don't want to be in the dark when a stalled table saw blows a breaker.

You also don't want to be in the hospital, so ground all sockets, along with all your metal tool cabinets and stands, to a separate grounding rod. You should never depend on the neutral wire in the 220 feeds to act as a ground: A fourth wire is essential to keep you from lighting up like a Christmas tree if a switch or motor should short out. Don't wire the shop without inspection or certification by a licensed electrician.

Also make sure that there is an adequate lighting system. Fluorescent lighting is an excellent main source of light for the central work area; four to six 8-ft. long double strips provide a pervasive, relatively shadow-free light. But because I dislike the sterile, pulsating quality of this light, I mix in the steady warm light of incandescents, usually placing a 300-watt bulb on the ceiling toward each corner of the shop. There also exists a need for spot illumination at individual work stations throughout the shop, especially on the drill press, the bandsaw and the sharpening/grinding area. These spots should be on individual switches so they can remain off except when needed. In all cases, light sources should be located as high as possible in the space to reduce glare.

Heating

The only source of energy I have ever used for heating my shop is, wouldn't you know it, wood, and I'll bet most small, custom woodworking shops use their raw material for heat. For cabinetmakers, kindling is plentiful (and we already know how to get splinters out). Beyond that, wood is still a cheaper source of Btus than any other fuel, and it allows the option of expending your own time and energy to obtain it.

But you do have to be careful. Woodstoves must be properly installed, and each type has a particular combustible-surface offset requirement. Don't forget that the area under the stove must be noncombustible as well. Exhaust pipes also have offset requirements, and they need insulated fittings where they pass through the structure. Check with your local fire department for information on the proper installation of a woodstove and exhaust system.

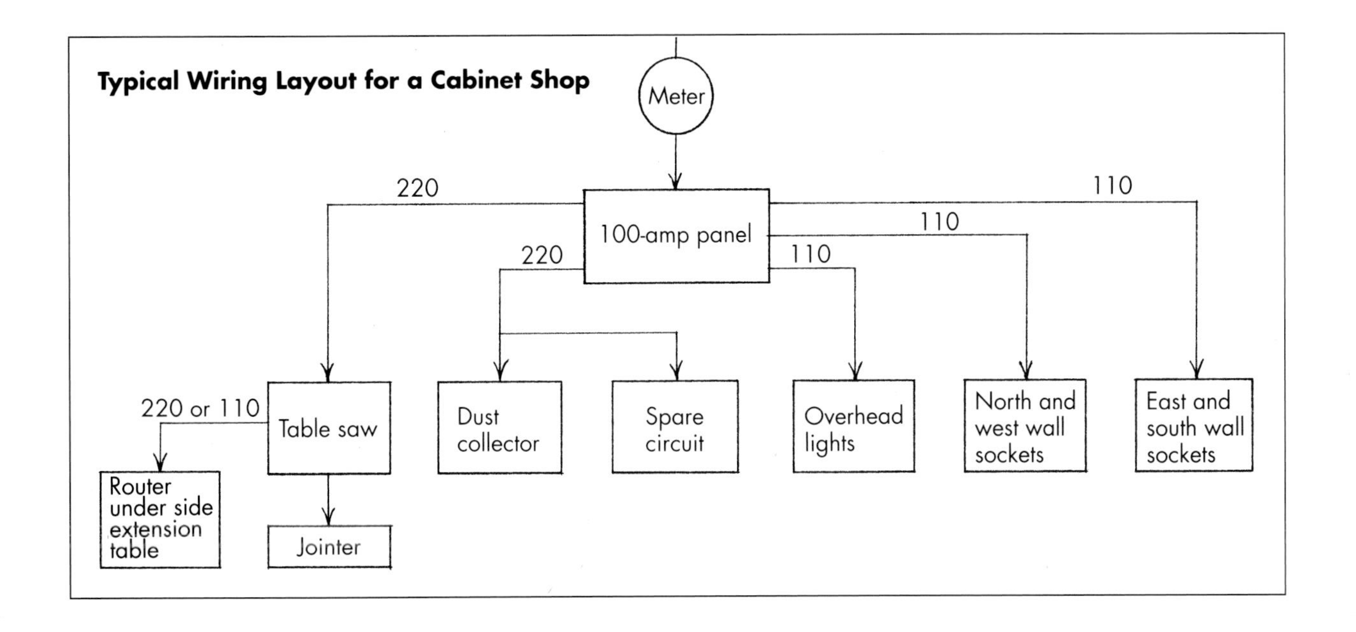

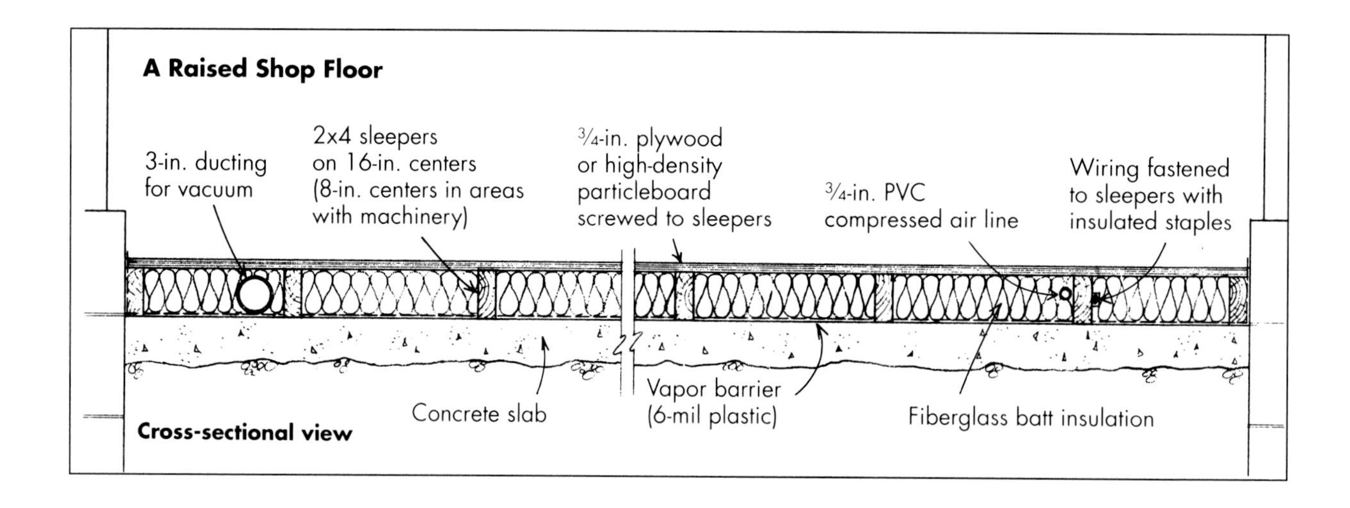

There are several other precautions peculiar to woodstoves in woodshops. If there's one thing that there is too much of in a woodshop, it's wood. Scraps of solid stock, plywood offcuts and rejected components build up quickly in the midst of production. Constant vigilance is necessary to keep this material contained in scrap bins. A piece of dry wood inadvertently left leaning on a wall within the combustion zone will catch on fire. If you're not there within 22 seconds of ignition, the fire will spread to the structure. If you're not there within 2 to 3 minutes, you have just lost your shop.

Also be picky about the type of scraps you burn. There is some dispute about the toxicity of the released gas, but I would avoid burning glue-laden materials such as particleboard, plywood and, especially, plastic-laminated sheet stock. Immerse oily rags in water, don't burn them in the stove.

There are two amenities that can make the use of woodstoves more efficient. Ceiling-mounted fans force warm air down to the floor, allowing the shop to heat up quickly and evenly. A woodbox, preferably one that can be filled from outside the shop, contains the dirt and those nasty bugs that eat wood faster than a cabinetmaker ripping stock without a dust mask.

Plumbing

The ideal shop will have a little addition that houses the office and a bathroom with a toilet and sink. Of course, if perfection is elusive, you can get by with a portable toilet. Running water is a pretty big plus for a woodworking shop, however; aside from the obvious hygienic benefits, soap and water keep hands clean and thus protect raw wood from finish-marring stains. Water also serves as a solvent for various finishes and for use with some sharpening stones. And there is nothing like a splash of cold water on the face to keep you alert through hot, dusty production runs.

Structural Modifications

Besides the possible addition of a small office space, there may also exist a need to add or modify the shop's doors and windows. It's important, for example, to have a large door, preferably a sliding type, located directly ahead of the table saw. As you will see in the discussion of machine layout in Chapter 2, the major stationary tools are grouped in as compact a space as possible. A door ahead of the table saw allows this tool group to be positioned to within 8 ft. of the shop wall; opening the door allows the ripping and jointing of any stock longer than 8 ft. If for some reason it's not possible to add a door (you have a recalcitrant landlord, for example), perhaps you can get away with a sliding panel in the wall that is just large enough to pass stock through.

You'll need a second door through which raw materials can enter the shop and completed cabinetry can be removed. It should be as large as possible. My shop has large bypass doors on the access face. One slides back when I need room ahead of the saws and jointer; sliding back the other completely opens the wall ahead of the area where I stack cabinets after assembly.

If the windows of your shop are poorly placed and you are unable to move them, consider covering them with white nylon or canvas to eliminate glare. A good way to brighten the shop's interior is to paint walls and ceiling with a semi-gloss white paint. In one shop I rented while on a limited budget, I mixed up a large batch of traditional whitewash from 4½ gal. of milk, 50 lb. of hydrated lime and 2½ lb. of salt and sprayed it over every surface. It instantly eliminated grime and dark corners, and left behind an amazingly fresh smell.

Another alternative is to nail or screw white-faced building board, sometimes called insulation board, over the existing wall covering or framing. As well as brightening up the interior of the shop, the building board will add insulation and significantly lower noise. Another advantage is that it turns every wall into a bulletin board.

If your shop has a concrete slab for a floor, and most do, consider covering it with something more resilient. If you try to live with the concrete floor, your legs will ache and fatigue will be your constant afternoon companion. You'll ruin your back and every chisel you own, as they all will drop on the floor edge first in obeisance to the laws of physics (and always just after sharpening, in obeisance to the laws of Mr. Murphy). One solution is to cover the concrete with building board, simply laying the sheets snugly against one another. (You can buy the board with a black facing, which is less expensive than the white-faced type.) Don't glue the boards down or they'll be difficult to replace—you'll have to replace them in the heavy-traffic areas every few years.

A better, but more expensive and time-consuming solution is the floor construction shown in the drawing on the facing page. A raised, "soft" floor is created by laying down 2x4s on 16-in. centers over 6-mil thickness vapor barrier. Double the framing to 8-in. centers under the table saw/jointer group to eliminate any give in the floor in this area. Compensate for any gaps between the 2x4 sleepers and the slab, caused by hollows in the concrete, by inserting cedar shims. Fasten the sleepers to the concrete slab with powder-actuated fasteners, or alternatively by through-bolting to lead anchors set into the concrete. Run all the stationary tool wiring, compressed air lines and 3-in. vacuum ducting along the sides of the sleepers, insert fiberglass batting and screw the ¾-in. flooring in place. Just think of it: no more tripping over cords and hoses. To make sweeping easy, cover the flooring with several coats of deck enamel or polyurethane.

Chapter 2: Shop Layout

Let's assume you've found a shop space in a suitable location: It's 22 ft. by 24 ft. (the size of the ubiquitous two-car garage), with an additional shed extension running the length of one side, which you can convert into an office, bathroom and dead storage. You've brightened up the walls, added building board to the ceilings and installed additional lighting where necessary. A benevolent built-up floor, through which you've run wiring and ducting, lies underfoot. It looks like it's time to move in and go to work.

But before setting down any machinery, think about layout, because proper placement of the major stationary tools and work surfaces is critical to the creation of a shop that works for you. A good layout will maximize production, reduce operator fatigue and facilitate the smooth flow of materials throughout the shop. Shown on the facing page is the layout that I have found to be nearly ideal for a space of this size. The clustered tool groups are readily apparent in the drawing. Note how this arrangement maximizes the amount of open floor space. Raw materials enter the shop through the sliding doors on the west wall and are stacked in the north section of the shop on the multipurpose work platform or on lumber racks above the radial-arm-saw table. After processing on the saw/jointer/router tool group, the sized materials are stacked against the east wall. After further processing at the drill-press station or on the work platform (which will have to be reoriented), the components flow to the southern portion of the shop to await final assembly. Give this layout a fair trial. If necessary, you can always modify it to suit your idiosyncrasies.

The Saw/Jointer/Router Group

The grouping of the table saw (with a 3-hp plunge router mounted under the side extension table to serve as a shaper), jointer and radial-arm saw comprises the very heart of the cabinet shop. It is worth every effort to ensure that all are securely fixed in proper orientation to each other.

The drawing on p. 10 depicts this section of the shop. You can see that the table saw, router and radial-arm-saw tables, as well as the wood fence bolted to the jointer fence, are at exactly the same leveled height. This allows all the support systems to augment each other: The radial-arm-saw table doubles as a support for the router; the jointer fence doubles as a support when cutting wide stock on the table saw. It is this system of tables and supports that makes it possible for one person to maneuver full sheets of plywood and long lengths of solid stock single-handedly.

As you would expect, there is a lot going on in this area—it is the most important, most frequently used work station in the shop. Let's examine some of the details.

In the drawing on p. 10, note the drawers built directly under the radial-arm-saw table. These drawers can con-

Shop Layout for a Typical Two-Car Garage

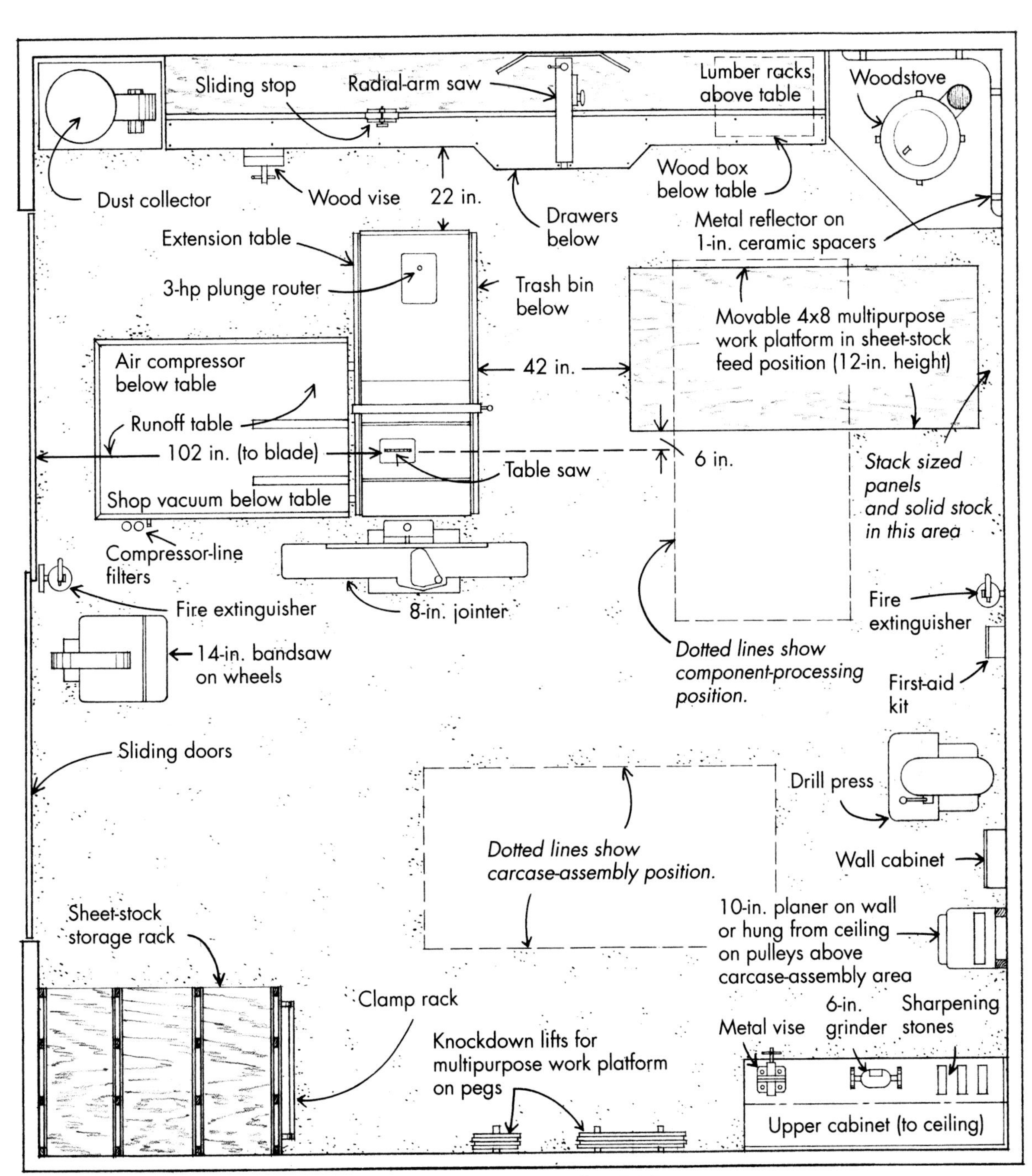

Shop dimensions: 22 ft. x 24 ft.

The northwest corner of Tolpin's shop houses the major tools—table saw with side extension table and underhanging router, 8-in. jointer with support fence, radial-arm saw and extension table, bandsaw and dust collector. Note the bin for wood scraps under the table saw's extension table and the work platform in position with sheet stock ready for milling.

tain extra blades, push sticks and measurement jigs for the saws; router bits and wrenches for the router; and any other miscellaneous accessories associated with this tool group.

The area beneath the router table houses a wheeled scrap bin (see p. 33), which receives cutoffs from both saws. The runoff table ahead of the table saw, shown in the drawing on p. 9, is built as large as space permits, to allow room below the table for an air compressor and mobile shop vacuum.

Finally, in the same drawing, note the addition of a wood vise to the far left of the radial-arm-saw table. The vise allows this table to double as a planing bench for hand-edging or shaping long stock, eliminating the need for a second bench. (Other vise functions are handled by knockdown clamp-support systems, as shown in the photo on p. 72.)

The Drill-Press Station

The drill-press station is another busy area in the small cabinet shop, especially if multiple attachments are used to perform certain processes in the 32mm system. Details on the 32mm system are given in Chapter 12. In brief, the drill press is set up to drill the holes on door backs for European-style cup hinges;

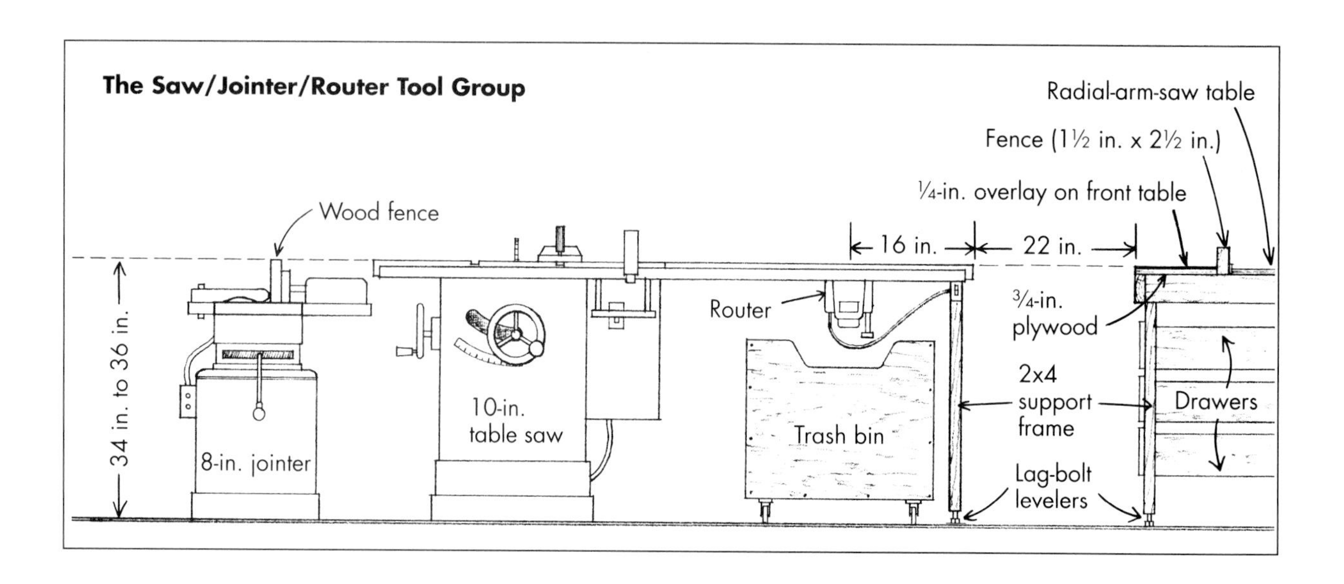

with the addition of a second attachment, it can simultaneously drill up to seven 5mm "system holes" in case components. These holes, which are spaced 32mm apart, receive hardware such as hinge-mounting plates, drawer slides and adjustable shelf clips.

Whatever it is used for, the drill press needs to be located centrally within the production flow—between the area where sized panels and solid stock are stacked and the component-assembly area. (Refer to the layout plan on p. 9.) This station also requires a wall cabinet within reach of the quill for storage of drill bits and other related accessories. One other suggestion: Chain the chuck key to some part of the machine. If you don't, you shouldn't complain when the key goes to school in your six-year-old's lunchbox someday.

The Grinding/Sharpening Area

You'll notice in the layout plan on p. 9 that the grinding/sharpening station is in the corner diagonally opposite the saw/jointer group, where it is out of the way yet readily accessible during most production processes. This is the one area where a little floor space can be sacrificed for a fixed cabinet. It will contain a host of items, from hardware and sharpening and grinding accessories to hand or power tools not housed in the mobile tool caddy (see p. 32). Locate a wall cabinet for additional tools, fasteners and finishing supplies above the bench. Build it right to the ceiling and provide access with a folding stepladder.

In addition to a grinder/sander unit and your sharpening stone or wet-wheel system, the bench can also house a metal vise and perhaps even a small bench-mounted drill press, which comes in handy when the primary unit is tied up in production.

The southeast corner of the shop contains the drill press, grinding/sharpening station and storage cabinets. A rolling tool cart is visible in the foreground.

The Placement of Mobile Tools

I'm going to make two assumptions here: first, that your dust collector is immobilized in a corner of the shop to which you've run ductwork, and second, that you are willing to forego a shaper. I have found that nearly all my shaping operations can be performed on the 3-hp plunge router mounted under a table. Occasionally I will use molding heads on the radial-arm saw or table saw to shape large patterns, and I have even farmed out volume work to mill shops. In a shop of this size, an additional stationary tool creates a real space burden with little net gain in productivity. This said, there really are only two major mobile tools to deal with: the bandsaw and the surface planer.

The bandsaw has spent a good deal of time roaming around my shop. Its latest location is just ahead and to the left of the jointer (as shown in the drawing on p. 9). So far, its presence here hasn't interfered with the saw/jointer/router group, and I rarely have to move it (it's on casters) to use it. Maybe it's finally found a home.

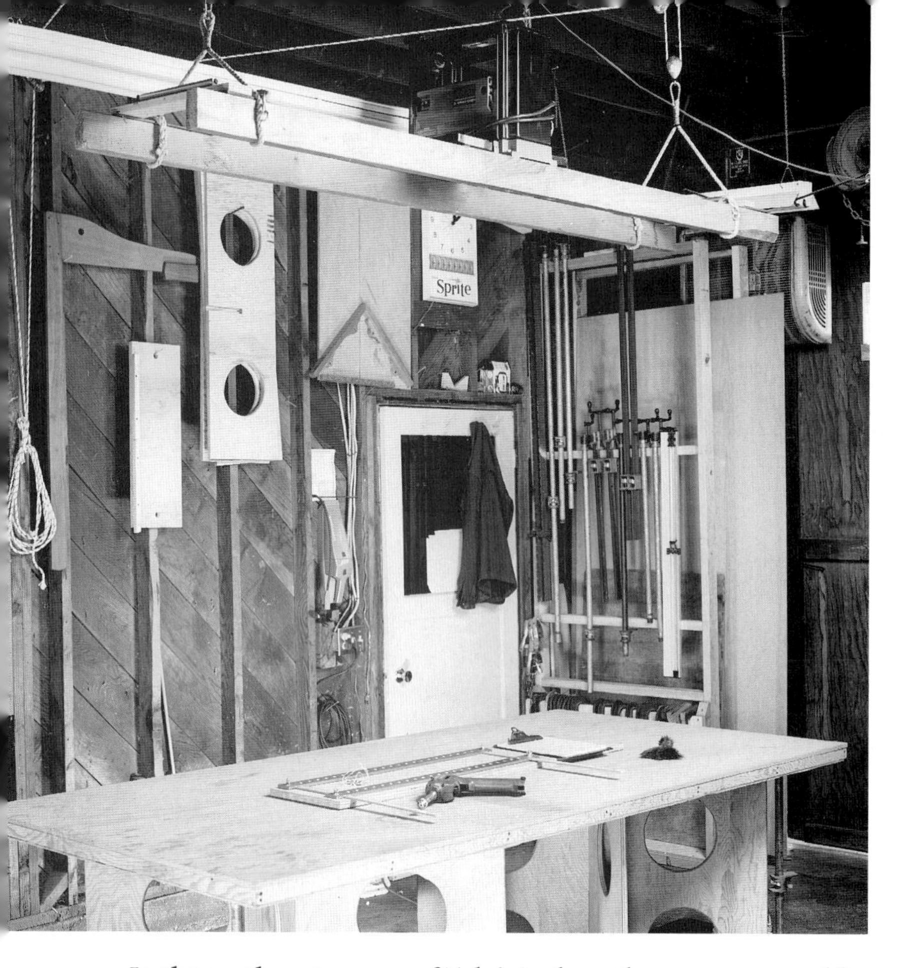

In the southwest corner of Tolpin's shop, the carcase-assembly platform sits on plywood lifts. The rack in the corner is for stacking surplus sheet stock; it doubles as a clamp rack. The wall provides storage for various pieces of equipment. Note the 10-in. surface planer hung from the ceiling. The grid support for the planer is shown in detail in the drawing below.

You may have been surprised to see the surface planer included as a mobile tool. But since the advent of the new pocket planers, such as the Ryobi 10-in. and the Makita 12-in., I have discovered that I no longer need the services of the space-eating, cast-iron monsters of old. Perhaps the greatest asset of the new planers is their light weight, which allows them to be mounted on a 2x4 grid, complete with feed rollers, and hung up out of the way on the wall or ceiling of the shop when not in use (see the photo at left and the drawing below). Casters on the ends of the unit allow the tool to be maneuvered into position.

The Placement of Work Surfaces and Storage Systems

There are a number of inexpensive fixtures you can build to help life go significantly smoother in the workshop. All the construction details are presented in Chapter 4, so for now, let's talk about the proper placement of the fixtures within the layout.

If you'll refer again to the drawing on p. 9, you'll see the 4-ft. by 8-ft. multipurpose work platform positioned adjacent to the table saw. In this orientation,

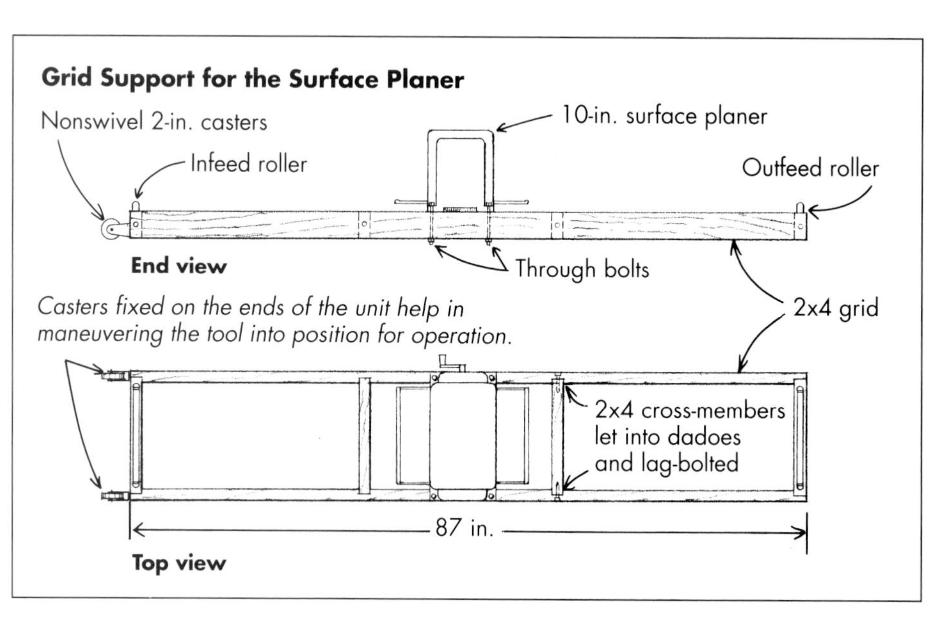

it serves as a sheet-stock feed support; it sits 12 in. above the floor on a pair of knockdown lifts. The sequence of photographs at right shows how one person can easily maneuver a full sheet of plywood through the table saw. When all the stock has been fed, the platform can be reoriented through 90° and elevated 32 in. to 34 in., whichever feels more comfortable, on another pair of lifts (indicated by dotted lines in the drawing on p. 9). In this position, the platform is a component-processing table. Later, the platform can move back down onto the 12-in. lifts, providing a carcase-assembly area (also shown in the drawing on p. 9).

Besides the work platform, you can also construct a number of storage racks. Locate a vertical storage rack for sheet stock in the last unused corner—diagonally opposite the woodstove. If it's possible to indent this rack into the dead storage area of the shed, or perhaps even to build a little shed off the shop just for it, so much the better. You can always use that extra floor space when you are assembling the cabinets.

The lumber rack can consume nearly the entire wall above the radial-arm-saw table. It is used primarily to store uncommitted stock out of the way of the production flow. Finally, locate the clamp rack on a wall as close as possible to the carcase-assembly area (as shown in the photo on the facing page).

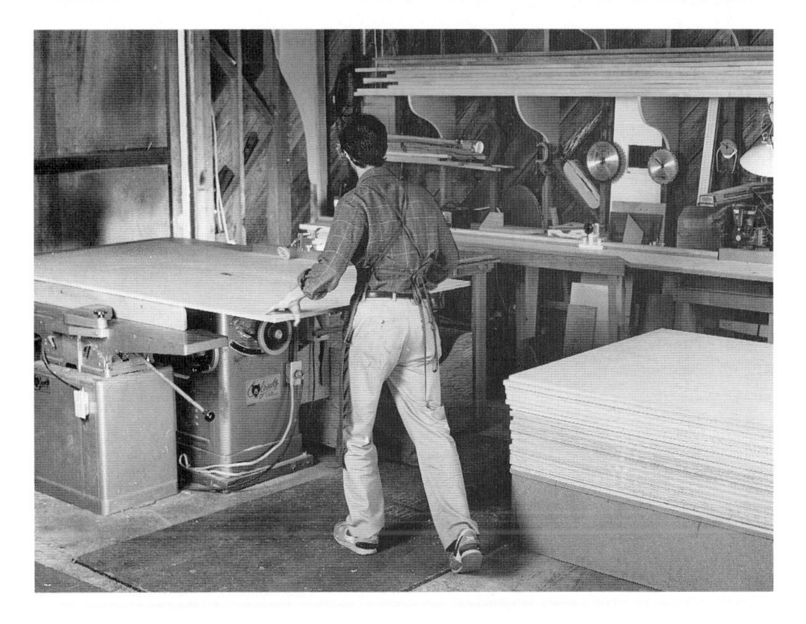

The first step in ripping a full 4x8 sheet of plywood is to lift its front edge and draw it up onto the table saw (top right). Then walk to the rear of the sheet and lift it level to the table-saw surface. Align the edge against the rip fence (middle right). Push the sheet through the saw (bottom right). The shop helpers assist in keeping the sheet indexed against the fence. The runoff table supports the sheet as it leaves the saw.

Chapter 3: The Tools of the Trade

Cabinetmaking, when practiced on the scale of the independent craftsman, is really not a complex industry. Efficiently made, high-quality products that reward the producer with a comfortable livelihood can be manufactured using the basic workhorses of the average woodworking studio, plus a decent collection of hand tools and hand-held power tools. This chapter, then, is a review of the tools that I have found to be essential for achieving success, and a collection of comments on adjusting them for maximum efficiency.

To make a living at woodworking, you need to own high-quality tools. As I'm sure you've already discovered, you get what you pay for: Good tools cost good money. Among power tools, a more expensive model will usually have a more durable motor and be smoother running and more comfortable to use. The Asian import machinery, which is still cheaper than its American prototypes, is something of an exception to the rule. Yet as of this writing, the quality, resale value and guarantee of endurance are not quite up to American-made standards.

Good hand tools, no matter where they come from, will cost considerably more than inferior tools. However, with some exceptions (especially if you are an aficionado of Japanese toolmakers or cottage industries), even the best tools currently on the market are rarely a match for the quality offered by the professional-grade tools manufactured in this country prior to World War II. It's worth seeking them out before spending the same amount of money, or more, on new tools. (See Appendix I on pp. 140-142 for a source for old tools.)

The Table Saw

As a cabinetmaker, you're going to spend a lot of time making plywood smaller than the way you bought it and making boards narrower than the way they came off the tree. The table saw will be your constant companion in these endeavors, so you want it to be the very best one you can afford to buy.

You can purchase a new car, and then some, for the price of certain table saws, but unless you are gearing up to build strictly European-style laminate cabinetry, you won't need a multi-bladed, panel-carriaged, computer-driven monster lurking in the shop. What you will need is a solid, relatively powerful, vibration-free, tilting-arbor saw with at least a 48-in. rip capacity and an accurately indexed fence system.

You can, in fact, get by with a 10-in. contractor's-type saw, and I did for many years. If your saw is equipped with extension and runoff tables (see Chapter 4) and an adequate fence, you can then upgrade other critical equipment before coming back to the table saw. I do, however, strongly recommend that you eventually acquire one of the heavier, enclosed-stand units epitomized by the Delta Unisaw. What you gain from a saw of this class is legion. Instead of the contractor's 1-hp or 1½-hp motor, you get 3-hp transferring power

The table saw's rip fence has slots that carry the shop helpers and allow them to be locked in various positions. (A fence that slides freely and can be set accurately at any length up to at least 48 in. to the right of the blade is essential.) The panic bar beneath the rail allows the operator to shut off the saw with his or her hip.

over three belts (not just one) to an arbor encased in a heavy, vibration-dampening iron casting. The motor is enclosed and therefore less susceptible to collecting dust. The switch will more than likely be magnetic, which can protect you from shock and automatically remains shut off after a power break. Finally, if you're accustomed to the controls of lesser saws, you'll really appreciate the beefy cranks that will easily and smoothly change the height of the blade and the angle of the arbor.

Obviously, for consistently accurate, smooth cuts, you want to make sure that your machine is correctly adjusted according to the owner's manual. (See Appendix II on p. 143 for other sources of information on fine-tuning of the various machines.) For some of these adjustments, I have found a dial indicator to be invaluable. This machinist's device has plenty of uses, and it's less expensive than a good hand drill.

Using the dial indicator, you can ascertain whether the flange against which the saw's blades are bolted is free of wobble—an important thing to know and a check often omitted from saw manufacturers' manuals. With the dial indicator fastened to the table and its arm set against the flange, hand-rotate the arbor and read the amount of variation (or hop-out). It should be less than .001 in. If it's more, significant vibration may be occurring at the cutting edge of the blade, which would prevent you from ever getting a smooth, splinter-free cut, especially in laminated sheet stock.

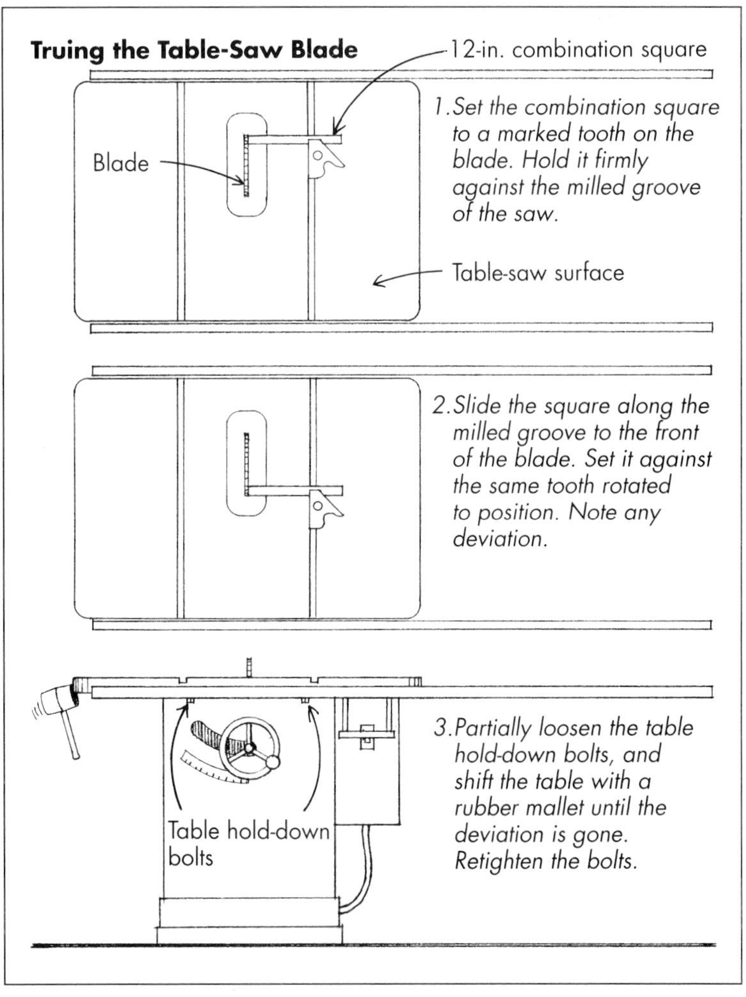

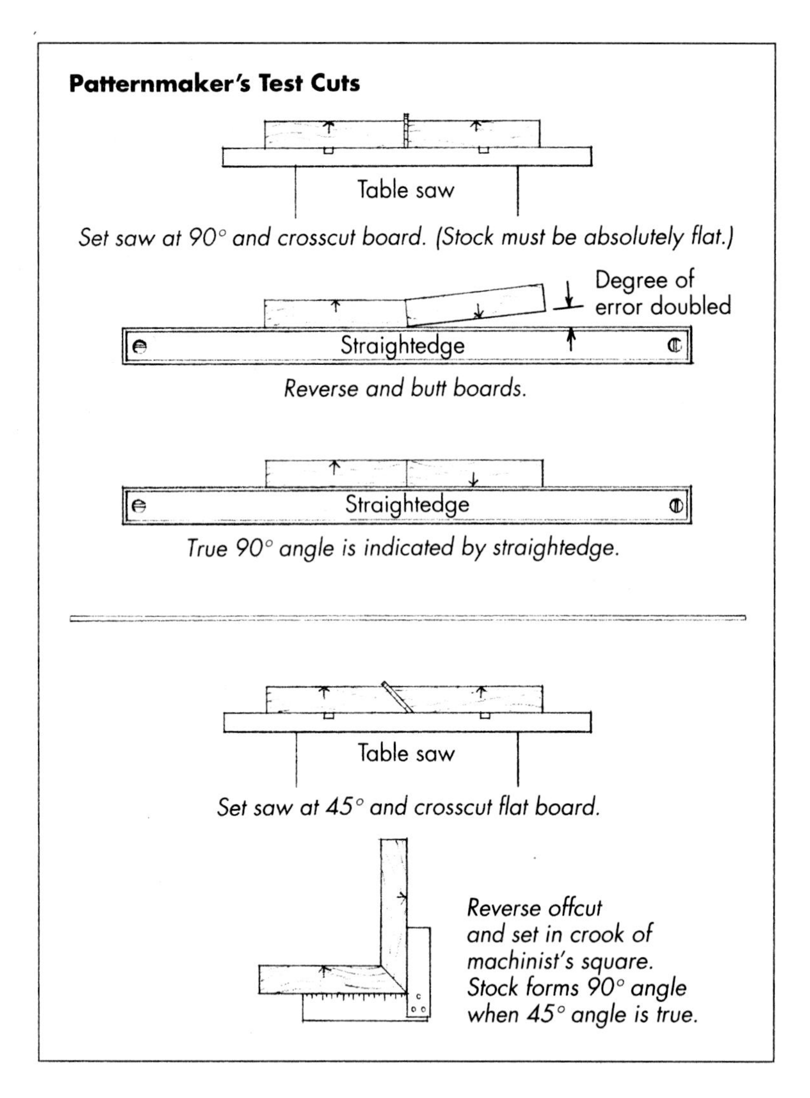

If you have a problem here, sigh deeply; the only solution is to remove the arbor from the machine and have the flange faced at a good machine shop.

If the flange passes inspection, use a 12-in. combination square to check that the blade runs absolutely parallel to the milled grooves in the table surface. Parallelism is essential for smooth cuts when using any jigs, including the rip fence, because the jigs are indexed to the grooves. A sequence of three steps is taken to true the blade, as shown in the drawing on p. 15. Be sure to unplug the saw. First, hold the combination square firmly against the side of the groove closest to the blade while sliding the rule out until it just touches the extreme tip of a marked tooth on the sawblade. Lock the square in position. Second, rotate the marked tooth to the front of the saw and slide the square forward to meet it. If the plane of the blade and the side of the groove are parallel, the rule will touch the marked tooth with no deviation from that of the first step. If there is deviation, take the third step: partially loosen the bolts that hold the table surface to the stand and strike a rubber mallet against the side of the table to nullify the deviation. The remaining friction in the bolts prevents the table from moving in large steps. After double-checking the back and front of the blade, retighten the bolts.

Also check the blade angle stop at 90° and 45°, using the patternmaker's trick of reversing test cuts to see whether the saw is true at these angles (see the drawing at left). At a true 90° angle the reversed and butted boards will lie flat against a straightedge. At a true 45° angle the boards when reassembled will form a perfect 90° angle. Consult the owner's manual to adjust and tighten down the lock bolts. Finally, make sure the rip fence locks down exactly parallel to the table grooves. Adjust it according to the manufacturer's instructions and lock securely. (I use a liquid lockwasher made by Duro to prevent adjustment nuts from loosening with vibration.) Make several test cuts and adjust the width indicator on the fence.

The accessories you select can add to or sabotage the versatility and safety of your table saw. As with the saw itself, with sawblades you get what you pay for. I recommend that you start with the following small selection of high-quality carbide blades: a 24-tooth rip blade, a general-purpose 50-tooth combination blade and a 60-tooth "triple-chip" grind-pattern blade for use with plywood and laminated sheet stock.

You need a good fence system on your table saw, one that locks accurately

The Excalibur fence features optional fixtures to hold a router fence in place. The fence can be adjusted in and out with a turn of a knob. Note that the shop helpers are slid in position to act as hold-downs for routing operations.

and securely at any distance along its track and that moves effortlessly from setting to setting. (See Appendix I on pp. 140-142 for suppliers.) With the fence, I recommend using the shop helpers made by Western Commercial Products (see Appendix I); these yellow rubber wheels hold stock firmly against the fence during ripping operations. Since they turn in only one direction, they also prevent kickback. I have found shop helpers to be especially useful when cutting large pieces of sheet stock. The fence system offered by Excalibur is unique in that it accommodates these hold-downs along a built-in track, as shown in the photo above. With optional equipment, the fence can also be used with a plunge-router setup. You should, however, provide for a second router fence for those operations where a molding pass on the router alternates with a rip cut on the saw. This second fence bolts to the router table through two slotted holes. (See the router-table construction details on pp. 26-27.)

Besides an effective hold-down system, other necessary safety accessories include a functioning blade-guard system and a splitter mounted just behind the blade to prevent wood from binding during ripping. Also note in the photo on p. 15 the bar mounted below the fence rail on the front of the saw. This device, which was designed and built for me by Jim Johnson, of Carlotta, Calif., allows me to turn off the saw with a simple bump of the hip. It's fast and certain, and it leaves my hands free to control the stock. Finally, I recommend waxing or otherwise lubricating (but not with oils) the surface of the saw, to prevent the wood from sticking and moving erratically through the blade. All of these accessories work together to prevent surprise—the last thing you want when working with a table saw.

The Radial-Arm Saw

When I began woodworking 20 years ago, I ran directly into the raging controversy of the time: "Which is better for the independent woodworker: the table saw or the radial-arm saw?" The table saw was indisputably king when it came to ripping stock and handling sheets of plywood, but the radial-arm saw was the champion of versatility and accurate production crosscutting. I solved the controversy for myself by eventually acquiring both and setting them up to complement each other's strengths.

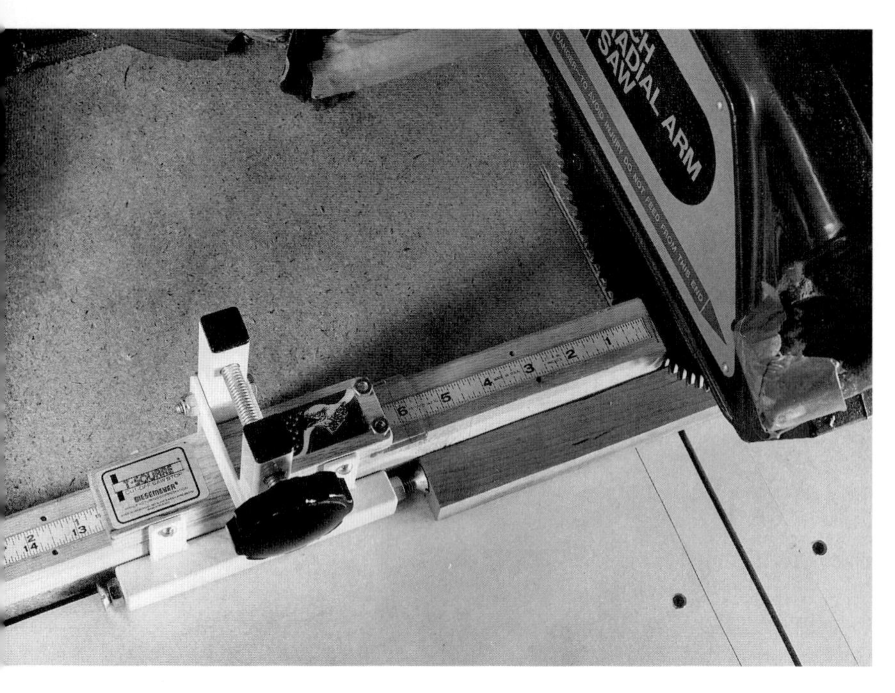

This is the Biesemeyer sliding stop that Tolpin uses on his radial-arm saw. A right-to-left-reading tape extends the full length of the left-hand extension table.

As the layout plan on p. 9 shows, in my shop the radial-arm saw lives just a step away from the table saw. It is set up to do just that task that it does better than any other tool: accurately crosscutting stock at 90°. Once the radial-arm saw is correctly adjusted, the settings are never changed except for occasional fine-tuning. If cuts are called for at angles other than 90°, wedges can be set against the back fence to bring the stock out to the desired angle. Mitering operations are not performed on the radial-arm saw; instead, a sliding jig is used for this purpose on the table saw.

When choosing a radial-arm saw, look for a machine that is sturdily built. It should have carriage-arm tracks of machined cast iron or replaceable steel rods; avoid saws that have tracks made from nonmilled sheet metal, because they will never provide consistently smooth, accurate, splinter-free crosscuts. The size of the motor, blade and table are not nearly as critical as the overall quality and sturdiness of construction. Even a small 8-in. blade model will usually have the capacity to cut 2x12 stock at 90°. See Appendix II on p. 143 for a review of available saws.

For consistently smooth, accurate cuts, some blade-alignment adjustments are necessary. The owner's manual should explain how to adjust for horizontal alignment, for squareness of cut relative to the fence and the table surface and for vertical alignment. This last adjustment will ensure that the front and back of the blade are perfectly aligned with the travel of the saw along the carriage arm. Thus a horizontal line drawn through the plane of the sawblade will parallel the line produced by the cutting action of the blade. Perfect alignment here is critical to producing warp-free and burn-free crosscuts.

Only one blade is required for the radial-arm saw as it is used in my shop: a high-quality, 60-tooth ATB (alternating top bevel) carbide crosscut blade. Although the exact number of teeth is not important, avoid extremely high numbers; I've found that these blades tend to bind in solid stock.

In my shop, the radial-arm saw is surrounded by extension tables designed to handle long lengths of solid stock. A dust-collection duct located just behind the blade removes nearly all the dust created during a crosscut.

The symbiotic relationship between the radial-arm saw and the table saw is consummated by the precise attunement of each machine's indexed fence or stop system. When the rip fence on the table saw cuts a piece of plywood at $24\frac{1}{32}$ in., the radial-arm saw, with an indexed stop set at $24\frac{1}{32}$ in., cuts a length of stock that precisely matches the width of the plywood. As shown in the photo above, I set up my radial-arm saw with the sliding stop and reverse-reading tape manufactured by Biesemeyer. (See Appendix I on pp. 140-142.) Once you have come to trust the relationship between the two machines, you'll find you can almost throw away your tape measure.

The Jointer

The jointer snuggles in right next to the table saw in my shop, an integral member of this tool grouping. As a rough edge is created on the table saw, the jointer is right there to straighten and smooth that edge in preparation for the next operation, be it a second pass through the rip blade, a crosscut at the radial-arm saw or a molding pass on the router. It all happens here in this one corner of the shop. The relationship between jointer and table saw is further cemented by the former receiving its 220-volt electrical connection directly from the latter's 220-volt feed. In return, the jointer offers a length of hardwood bolted to its fence as a secondary support when large sheets of plywood or long jigs must be maneuvered on the table saw (as shown in the sequence of photos on p. 13).

You can get by with a 6-in. jointer, as I did for many years, but when you can acquire an 8-in. machine, do so. You'll wonder how you ever managed with the smaller one. It's not so much the extra power or table width, but rather the longer bed that is such a blessing to the one-person shop. You'll find that jointing a 12-ft. 1x12 single-handedly is no longer a problem, and you can permanently retire your 24-in. hand plane (except for fun and exercise).

Although the jointer is a rather simple tool with few moving parts, there are several checks you need to perform to ensure that it will produce an absolutely flat and straight surface on a piece of wood. First, check the condition of the tables. A known straightedge (machine shops sometimes have offcuts with a true edge) is used to check for flatness and parallelism between the two tables. Detect any twist in the table surfaces with winding sticks. Inserting machinists' brass shims in the ways on which the tables slide, or sometimes simply retightening the gib screws, will usually correct sagging (nonparallel) tables. Major problems of warpage, concavity or convexity can be corrected only by resurfacing at a machine shop. Try jointing a number of boards and checking the results before pulling a table—the tables may not be as bad as they look.

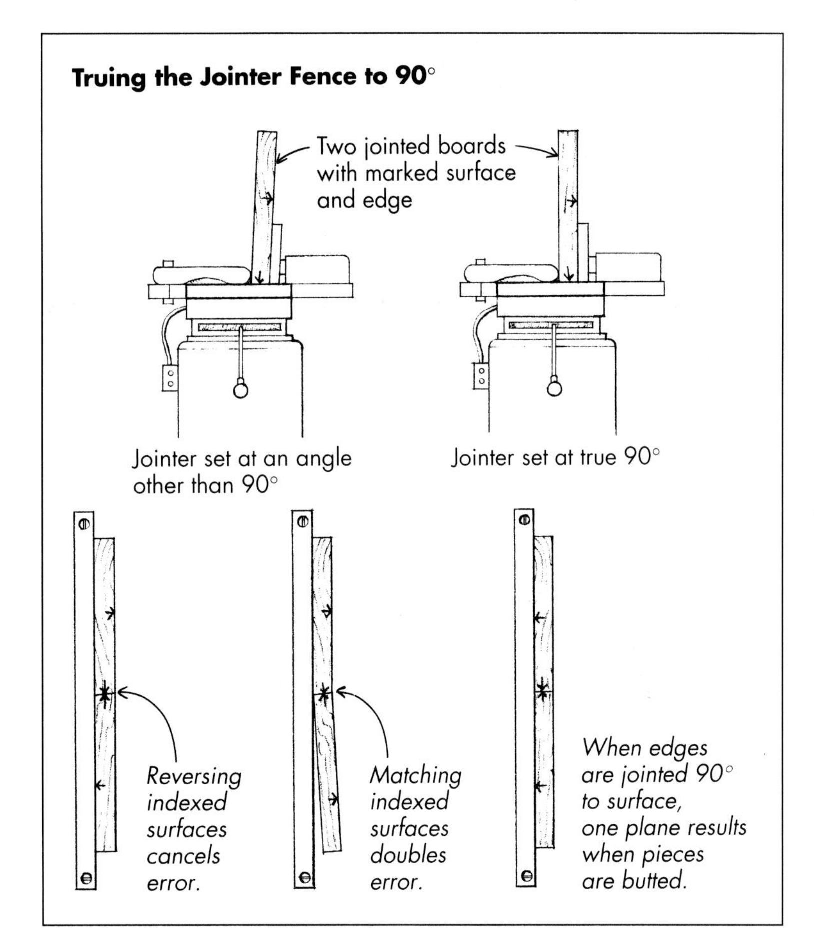

Adjust the jointer's fence, checking with your 6-in. machinist's square until it reads 90° at all points along the fence. The jointer's fence is trued to 90° by the comparison of two jointed boards. If any angle other than 90° has been planed on their edges, their indexed faces when oriented the same way will not form a straight line (as shown in the drawing above). Finally, make sure the blade guard is functioning properly. It should not bind when encountering stock and should close quickly behind the board as it passes. Wax the guard where it meets the wood and keep the pivot well

Here the planer has been lowered from the ceiling onto the assembly platform. A pair of 8-ft. long 2x4s placed crossways on the platform expands the platform's holding capacity.

lubricated with a light grease. Also lubricate table and fence surfaces with wax or a non-petroleum-type product to avoid erratic feed movement.

One final note: Save one-third of the blade area exclusively for edging stock that is damaging to the blades, such as particleboard and plywood. The remaining two-thirds will stay sharp far longer and will extend the blade replacement period significantly.

The Surface Planer

For many, many years I owned a handsome, old and heavy (actually, immobile) 16-in. planer. The motor itself was the size of today's typical 10-in. or 12-in. planer, weighed well over 100 lb. and was rated at 1 hp: That horse was one obese draft animal. The Red Fox (its brand name) commanded the middle of the shop and everyone worked around it. We had to, because it clearly wasn't going anywhere, at least not until the day the light came on in my dust-filled skull, and with the floor groaning a sigh of relief, I booted it out the door into the arms of an eager boatbuilder. (It actually took more than one boot and more than one boatbuilder.) In came its replacement: a modern, lightweight (50-lb.), 10-in. pocket planer made by Ryobi, which quickly found itself bolted to the 2x4 grid I had prepared for it (see the drawing on p. 12) and hung from the ceiling of the shop.

I was able to go with a light-duty planer after years of living with the Red Fox because my methodologies were changing. I was no longer buying wood in the rough, finding it far more cost-effective to let the mill surface it to within 1⁄16 in. of final dimension before it was delivered to my shop (stock sold this way is designated S2S, that is, surfaced two sides). By this time I was also farming out the surfacing of laminated panels to a local mill yard, which possessed an accurate, efficient, and, of course, expensive, surface sanding machine. Thus I required a surface planer to do not much more than the final, uniform dimensioning of stock and the occasional sizing of moldings, trim or spacers.

I won't spend much time discussing surface planers, as I find I can simply recommend you acquire one of the new lightweight ones. They are nearly as fast as much heavier machines, and because of their fast cutter speeds and extremely sharp knives (which are easy to keep sharp with the jigs supplied by the manufacturer), they produce a consistently smooth, nearly unblemished surface. They also appear to be quite durable when used for finish-surfacing and light milling, and you can hardly beat the price. Do go the extra mile when buying your planer and acquire the knife-sharpening jigs, a vacuum chute and infeed and outfeed rollers. However, if you simply can't live without *your* Red Fox, I suggest you build it a doghouse alongside the shop. You'll find something else to do with the vacated floor space, believe me.

The Drill Press

The drill-press station is a busy place in the production flow of a typical cabinet job. If face frames are called for, the drill press is the machine that will help you construct them swiftly, using the specialized drill bit discussed on p. 68. This station is also the place where you produce holes for adjustable drawer faces (see p. 76) and European-style cup hinges (see pp. 88-90). If you add a multi-spindle attachment, the drill press will also produce your 32mm-system holes (see p. 58). A variety of sanding and grinding attachments can greatly add to the versatility of the drill-press station.

Select a floor-standing model for your primary drill press, and securely bolt it there. It should have a speed range of at least 400 rpm to 2800 rpm and should shift easily from one speed to another. A 6-in. throat capacity, 3¼-in. spindle throw and ½-hp to ¾-hp motor will do the job. The table, to which you should add a replaceable surface (I simply use a 16-in. by 24-in. sheet of ¾-in. plywood), should tilt and lock securely at any angle. It should also sit exactly 90° to the travel of the quill when measured side to side, and from front to back of the table. The side-to-side measurement is adjusted by simply swiveling the table on its mount; the front-to-back measurement is adjusted by adding machinist's shim stock or fine sandpaper between the table and the plywood. Finally, a quick-setting, slippage-free stop (for depth regulation) is an absolute necessity for your drill press.

Like the jointer, the drill press is a rather simple machine—accurate performance depends on sturdy construction and close tolerances. Your trusty dial indicator can be used to check on a number of critical adjustments. Clamp its base to the table (which should be locked at 0°) and swing the indicator's arm against the drive spindle. When the spindle is rotated by hand, the dial will read the amount of hop-out. This should be less than .001 in. Continue by swinging the arm against the shank of a trued ½-in. steel rod (see your local machinist for a loaner) carefully centered in the chuck. The runout as measured against the bit should be less than .004 in.

Assuming that your drive spindle tolerance is within .001 in., a persistent wobble at the drill bit is probably traceable to an inferior chuck, which will not center properly. Replace it with a better chuck; if the problem persists, you may need to have the taper reground at a good machine shop.

The Bandsaw

It would be hard to live without my old 14-in. Rockwell bandsaw, although I probably could get by with a good jigsaw if I had to. But since the bandsaw lives on wheels and seems to have found an unobtrusive home in my shop, it doesn't have to worry about going the way of the Red Fox, a victim of my fanatical quest for unobstructed floor space. Actually, there really is nothing quite up to a bandsaw for cutting arches in the rails and panels of doors, and for introducing nice curves into various lintels, aprons and support brackets. I also use the saw for producing notches in the top of partitions and, occasionally, for roughing out a variety of hand-finished joints, such as bridles, half-laps and through-tenons.

A temporary rip fence and stop are set up on the bandsaw to aid in the notching of case components.

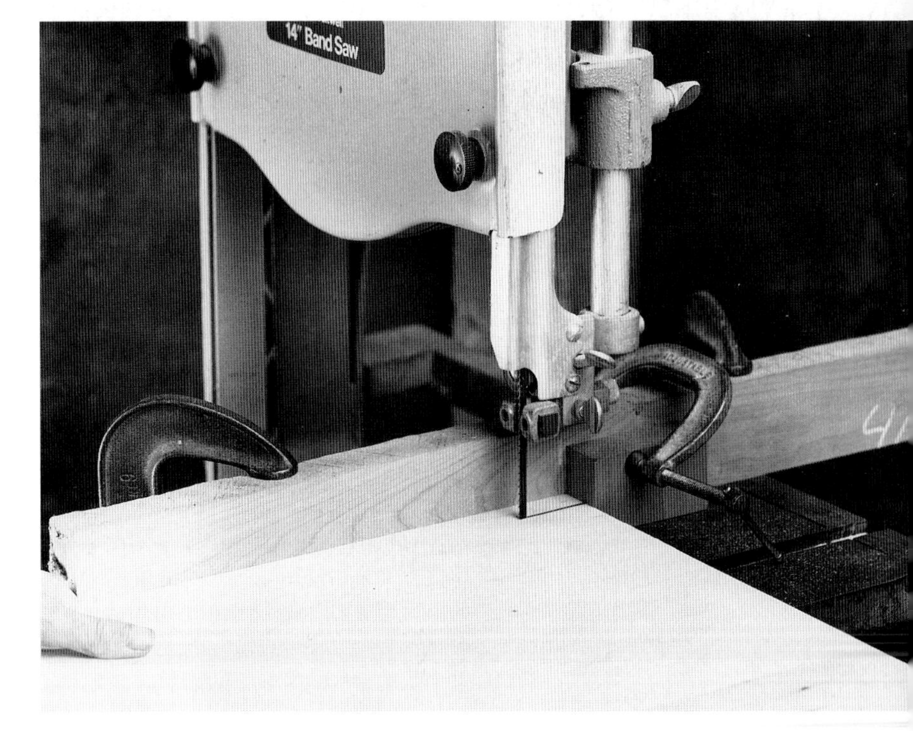

Unless you need resaw capabilities (perhaps you like to resaw found lumber for unusual door panels), you can get by with a ¾-hp saw with a 14-in. throat capacity and a 6-in. height of cut. You might even make do with a 12-in. saw, since its primary function will be to shape the edges of stock less than 2 in. thick. The saw should, however, be sturdily built. A good one will have two support carriages for the blade—one above and one below the table. A sealed ball bearing on each carriage supports the back of the blade when under pressure, and guide bushings support the sides. You will need to adjust the bearing carriages, guide bushings, tracking and tension according to the owner's manual. See Appendix II on p. 143 for a review of bandsaws and their maintenance and adjustment procedures.

The only blade you will probably ever need will be a ¼-in., 6-tooth-per-inch, skip-tooth blade, either of bimetal (which lasts a long time and can be highly tensioned for ripping) or of standard carbon steel. With either type of blade, relieve the tension when the tool will not be used for a long time. Other maintenance procedures include lubricating the table, cleaning the wheels of sawdust and keeping the motor pulley properly tensioned. Finally, a word about safety: Never run the saw with the blade guards, and especially the wheel covers, removed. I'll never forget the time we broke the blade on an ancient, 36-in. ship's saw. Without a guard in place over the top wheel, the blade—all 18 ft. of it—came slithering out like a snake, and with a blood-curdling shriek, it chased a bunch of boatbuilders around the shop and out into the rain. They wouldn't come back into the shop until the shop's welder had fashioned a cage for the wheel strong enough to contain a couple of impassioned gorillas.

The Air Compressor

An air compressor has many uses around a cabinet shop, and all are highly appreciated. Pressurized air blows wood chips and sawdust out of joints, tool motors and other hard-to-reach areas and cleans up surfaces for finishing. In addition, the compressor can power a wide variety of finishing, surfacing and drilling tools; it can also be used to blow up the tires on the truck and inflate the kid's dinosaur.

The first and foremost use of this tool around the shop, however, is to drive fasteners. There is absolutely no substitute for air-driven nail guns when you wish to fasten objects together rapidly, securely and without the fear of splitting out the wood or the need for predrilling pilot holes. Although the guns and the fasteners themselves are relatively expensive (and you should avoid trying to save money by buying inferior brands, because you won't), there are few tools in cabinetmaking that will pay for themselves faster than an air-nailing system.

There are a multitude of guns and fasteners, but you can get by with just two tools: a finish nailer with a capacity for finish nails up to 2 in. long, and a brad tacker for shooting ⅝-in. to 1-in. slight-headed brads. With these two guns, you can assemble drawers and other components; apply face frames, moldings, trim and bracing elements; aid in some assemblies; and apply on-site moldings and trim.

To power these tools, you'll need at least a ¾-hp, single-stage compressor unit with a minimum tank capacity of 4 gal. (Going with a larger motor and greater storage capacity will allow the machine to cycle less frequently, extending the life of the motor and your ears.) For shooting nails, the compressor should deliver at least 90 psi at 1.8 cfm (cubic feet per minute) through a regulator. If you intend to use the compressor to spray paint or lacquer, be sure it will deliver 5 cfm to 7 cfm at 30 psi to 38 psi; to prevent contamination of the finish materials, install oil and water traps (even if the compressor is an oilless type) at the compressor, or, ideally, near the end of the feed line.

Maintenance of an air compressor is relatively straightforward. Drain the tank on a regular basis to prevent moisture buildup and subsequent corrosion; periodically check the oil level in the crankcase; keep proper tension on the drive belt (if the machine has one); periodically check the action of the safety valve on the tank; and periodically drain the oil and water filters. For safety, make sure all moving parts are covered by a guard (compressors start on their own without warning), and periodically check the condition of the flexible hoses, especially in the vicinity of any fittings (spend your money on good hoses). Finally, a word of caution: Keep your skin away from close contact with high-volume, narrow-orificed compressed air. The air could get under your skin and into your bloodstream, and cause a vapor lock in your heart.

The Dust Collector

After more than a decade of eating dust and sliding on shavings, not only would I not live without a dust collector ever again, I probably *couldn't* live without one. If I had to do it all over, I would have bought a collection system the day I bought my first table saw, if, of course, they had made small-scale dust collectors back then. Fortunately, they do now: See Appendix II on p. 143 for an extensive review.

In a small, one-person shop, almost any of the models currently on the market will be able to suck away the waste of any one machine. The most voluminous producer is probably the surface planer, which, assuming a short length of ducting, will require about a 400-cfm draw from the dust collector to keep up. Even small machines with a 1-hp motor are generally rated above this level. If, however, you've decided to run ductwork through the shop as discussed on p. 7, I'd recommend at least a 2-hp model with a rating of better than 1000 cfm to make up for friction loss in the ducting and to allow drawing from more than one source at a time (for example, the jointer and the table saw in a sequential operation). You'll be installing gates at each station, tight-fitting to avoid vacuum loss, and switches to make activation of the collector convenient.

There are a number of safety considerations with this tool. If plastic ducting is used, a ground wire needs to be run through the pipe to eliminate static-charge buildup, which has been known to ignite fine dust. Metal ductwork should be grounded at some point as well. Also check to be sure that the impeller on the motor is nonferrous in composition (cast aluminum or plastic). Be sure the dust bags are securely fastened to their attachment points, even if a modification such as adding additional band clamps is necessary—if a blow-off occurs when full capacity is reached, the whole purpose of the system is defeated, at least for the rest of that day. Finally, keep anything you might ever want to see again away from the inlets to the system, including small lengths of molding, fastenings, router bits and pets. You haven't seen a dust-covered golden retriever with a piece of ogee molding in his teeth, have you?

Hand Tools

It's been said that the tools of man are but an extension of his wish to manipulate the world around him. I sense, then, that it is the hand tools, especially those driven by our muscles, that come closest to expressing the true nature of our desires. Unfortunately, in that we have so many desires, we also feel the need to have lots of tools. Try as we may, we are condemned to the feeling that we do not own the complete set. It's not such a problem, really (depending on whom you talk to); you just have to find or build the proper places to put away what you do have.

POWER HAND TOOLS

Belt sander: For hand-surfacing panels, face frames, etc., 3x21 or 24 is adequate; 4-in. width is better. Use the dust-bag attachment.

Orbital sanders: One each: ½-sheet and ¼-sheet capacities. I replace stock bases with pads manufactured by 3M for their "Stikit" sandpaper rolls.

Chopsaw: In my opinion, a dangerous and overrated tool, but necessary for on-site installation of trim and molding stock. Sliding miter jigs on the table saw and the radial-arm saw do the work of the chopsaw in the shop.

Circular saw: For occasional rough-cutting of stock to length, miscellaneous shop repair and construction, and inside cuts on sheet stock. A 7¼-in. capacity is adequate.

Jigsaw: Buy a good one. Look for adjustable reciprocating action and variable speed; tilting base a plus.

Laminate trimmer: For laminate installations, breaking edges on wood and producing small molding details.

Plunge router: 3 hp; permanently mounted under table saw's side extension table.

Standard router: For freehand use. 1 hp to 1¾ hp, can be plunge-type. Look for ½-in. collet capacity, positive locking mechanism and comfortable grips.

Pneumatic brad tacker: For assembling drawers and other components, attaching edge trim and applying various smaller moldings. Brad capacity should be ⅝ in. to 1 in. Buy a good one.

Pneumatic finish nailer: For attaching face frames, backs, larger trim and molding stock. Nail capacity should be 1 in. to 2 in. Better brands are more reliable.

Power screwdriver: Even a cheap one will make you feel like never driving another screw by hand.

Spline-biscuit cutter: The tool of choice for quickly creating accurate, strong joints in carcase components and solid-wood frames. Cheaper models are adequate.

Stationary spline-biscuit cutter: A step up from the previous item. Foot pedal controls feed of cutter into stock, freeing hands to manipulate stock. A low-overhead, high-production machine for face and door frames.

Variable-speed drills: Buy two, one a 9.6-volt cordless. Don't skimp on quality or you'll end up buying replacements. Look for ⅜-in. capacity, reverse and smooth speed transitions.

The tables at left and on the facing page list tools, grouped by power source, that I feel are essential to the efficient construction of cabinetry and casework-style furniture. You can find these tools at any of the mail-order houses listed in Appendix I on pp. 140-142.

The Sharpening Station

The sharpening-station bench contains all the equipment necessary to handle the shop's sharpening needs, with the exception of sharpening carbide-tipped circular-saw blades and router bits. A 6-in., ½-hp grinder sits near the middle of the left front corner, and a collection of waterstones and their holders sits to the right. Stored in the top drawer of the cabinet below are various sharpening jigs and accessories, a high-speed Dremel grinder-polisher and a small assortment of burnishing tools.

The shaft on one end of the 6-in. grinder is fitted with a ¾-in. by 6-in., 100-grit friable abrasive wheel. This type of wheel (which is available from Woodworker's Supply; see Appendix I on pp. 140-142) runs cooler than standard, nonfriable wheels and is used for the initial grinding of the primary bevels of drill bits, plane irons, chisels and other edge tools. A grinding jig, made by Veritas (also from Woodworker's Supply), ensures consistent alignment. The other end of the grinder's shaft swings a cotton buffing wheel impregnated with buffing compounds (available from jeweler-supply houses) for polishing knife-bladed tools (including some carving tools) to razor sharpness.

The razor edge on plane irons and chisels is not developed on a buffing wheel. For these tools, I create a more durable edge by developing a secondary bevel (or microbevel). I use a honing guide made by Eclipse (Veritas also makes a good one) to fix the blade securely over a sequence of Japanese waterstones of 800, 1500 and 6000 grit. One of the beauties of waterstones is that they are lubricated with water, avoiding the contagion of oil that can seriously damage many finishes. Perhaps an even greater benefit is the ability to reflatten waterstones by simply hand-rubbing them over a hard, flat surface. For this purpose I use a special mild-steel plate milled absolutely flat, but a ½-in. sheet of plate glass will also serve. The addition of a fine silicon-carbide powder (400 grit or higher) speeds the process considerably. It takes a perfectly flat stone to produce a perfectly flat, perfectly sharp edge.

I will occasionally fit planer knives from the jointer or surfacer into specialized jigs made for this purpose and hand-hone them over the waterstones. After the third honing I send them out with the carbide blades and install the fresh second set I keep on hand.

The metal vise mounted on the bench holds specialized drill bits (such as Forstner bits), plug cutters and high-speed-steel router bits in position for sharpening. I use either the Dremel tool fitted with a grinding or buffing wheel, or a set of small tapered files to bring these bits up to par. The vise also holds cabinet scrapers securely while creating the cutting burr with a burnisher.

HAND-POWERED TOOLS

These tools, stored in the mobile tool caddy (see p. 32), accompany me to each work station in the shop.

Marking equipment: Pencils (No. 2½) and sharpener, chalk, pens, marking knife, awl, mortise gauge and marking gauge.

Layout devices: Tape measure (¾ in. by 16 ft.). Squares: 24-in., 12-in. and 6-in. combination, 6-in. machinist's. Architect's scale, bevel gauge, protractor, dividers, calculator, trammel points and inside/outside calipers.

Cutting tools: Set of cabinetmaker's beveled-edge chisels (¼ in. to 2 in.), 1½-in. Japanese push chisel. Planes: block, finger, rabbet, No. 4 and No. 4½ smooth. Saws (all Japanese): dovetail, combination rip/crosscut, panel and keyhole. Utility knives.

Finishing tools: Set of cabinet scrapers, adjustable angle scraper holder, set of hand files, and glue scrapers in several shapes.

Fastening aids and devices: Set of Phillips-, slotted- and square-head screwdrivers. Set of high-speed drill bits. Hammers: 13-oz. claw, tack hammer and rubber mallet. Brad pusher, nail sets and punches. Set of Fuller countersink bits.

Grasping devices: Pliers with wire cutters, vise grips, end snippers, set of Allen wrenches, small set of open-end wrenches, nail-pulling lever and tack and staple remover.

This next set of tools can be found elsewhere in the shop, hanging on walls or stored in drawers below the sharpening-station bench. The hordes of router bits, specialized drill bits and other such accessories are not mentioned here—you'll meet them when we call for their help in Section II of this book.

Clamps: Four sets each of 2-ft., 3-ft. and 4-ft. pipe or bar clamps, numerous extra lengths of pipe. One dozen 6-in. C-clamps. Two 36-in. back-to-back bench clamps (made by Griset Industries). Four cam-activated guitar clamps. Several sizes of spring clamps, deep-throat clamps and band clamps.

Levels: Reservoir-type water level (for establishing reference line during installations), 30-in. and 78-in. bubble levels, torpedo level.

Miscellaneous: Glue applicators, roller for laying down laminates, electronic stud finder, putty knives in various shapes, spokeshaves, stiff-bristle file cleaner, layout ruler for 32mm-system machine setups, dial indicator, time clock, fire extinguishers and a good radio tuned to your favorite station.

Chapter 4: Shop-Built Fixtures

In this chapter, we will discuss a variety of homemade fixtures that will help you use your shop more efficiently. These include runoff and extension tables for the table saw and radial-arm saw, which fully support the materials being processed and thereby eliminate the need for a helper (even when handling full sheets of ¾-in. thick plywood). Also described is a multipurpose, multilevel work platform, which serves as a feed platform for sheet stock, a work surface on which components are processed and a carcase-assembly platform. Directions are given for building a variety of storage fixtures, which help to ensure that certain tools and materials get put away. The final project described is the construction of three mobile caddies, or "stooges." These caddies will be your constant companions around the shop. Moe carries the tools listed in the table on p. 25, Larry shoulders raw materials and component parts and Curley is a trash bin on wheels.

Table-Saw Extension and Runoff Tables

The extension and runoff tables provide flat, slippery surfaces precisely leveled to the plane of the table-saw table. The extension table (see the drawing below) completely fills the open space between the rip-fence guide rails extending to the right of the saw. The runoff table (see the drawing on the facing page) sits ahead of the saw and can be built as large as space permits. Both tables are

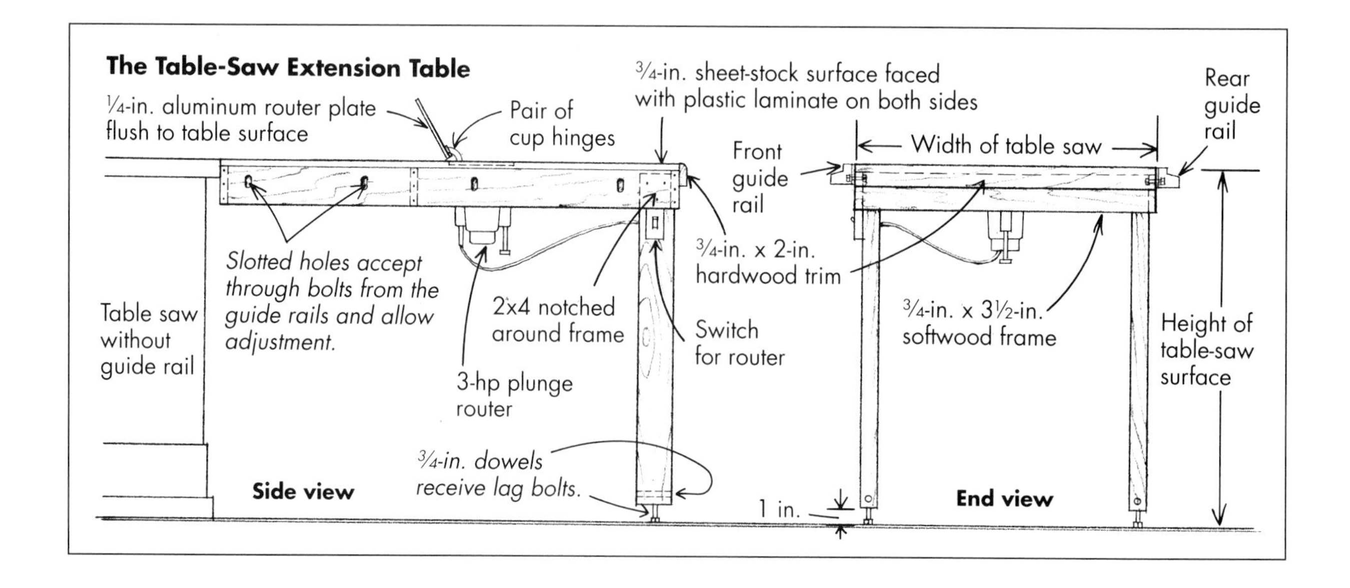

constructed of laminate-covered ¾-in. thick sheet stock screwed to a pine framework and supported on 2x4 legs with a built-in leveling system.

The first step in constructing the tables is to determine their dimensions. The extension table should fit the space between the guide rails precisely, while the runoff table should be sized so that it doesn't interfere with the operation of the jointer (or sliding table, if you have one). The runoff table should also allow a person to pass between it and the door ahead of the saw. When cutting the sheet stock, allow for the ¾-in. by 1½-in. hardwood end cap shown in the drawing at right.

Laminate the sheet stock on both sides to prevent uneven absorption of moisture and subsequent warpage. Select a neutral color with a matte finish to reduce glare—I chose battleship grey.

Build the support framework from ¾-in. by 3½-in. pine or alder, spline-biscuit-joining, gluing and screwing it together. Fasten the sheet surface to the softwood framing from below, using screws driven at an angle or figure-8 fasteners. The legs are constructed from 2x4s cut to length, notched around the frames and screwed in place. Inserted across the 2x4s near their lower ends are ¾-in. dowels. These receive the threads of lag bolts, which act as precision levelers when adjusting the surfaces flush to the table-saw surface.

The extension table is secured to the table saw by bolting the guide rails to the softwood framework. Make the shank holes in the frame elongated slots, so that it will be possible to adjust the height of the table. Note that the ¼-in. aluminum plate to which the 3-hp plunge router is attached is let in flush to the surface and hinged from below with cup hinges to allow access to the router from above. (The router plate with hinges is available from Excalibur; see Appendix I on pp. 140-142 for the address.) Also note the location of the switch for the plunge router.

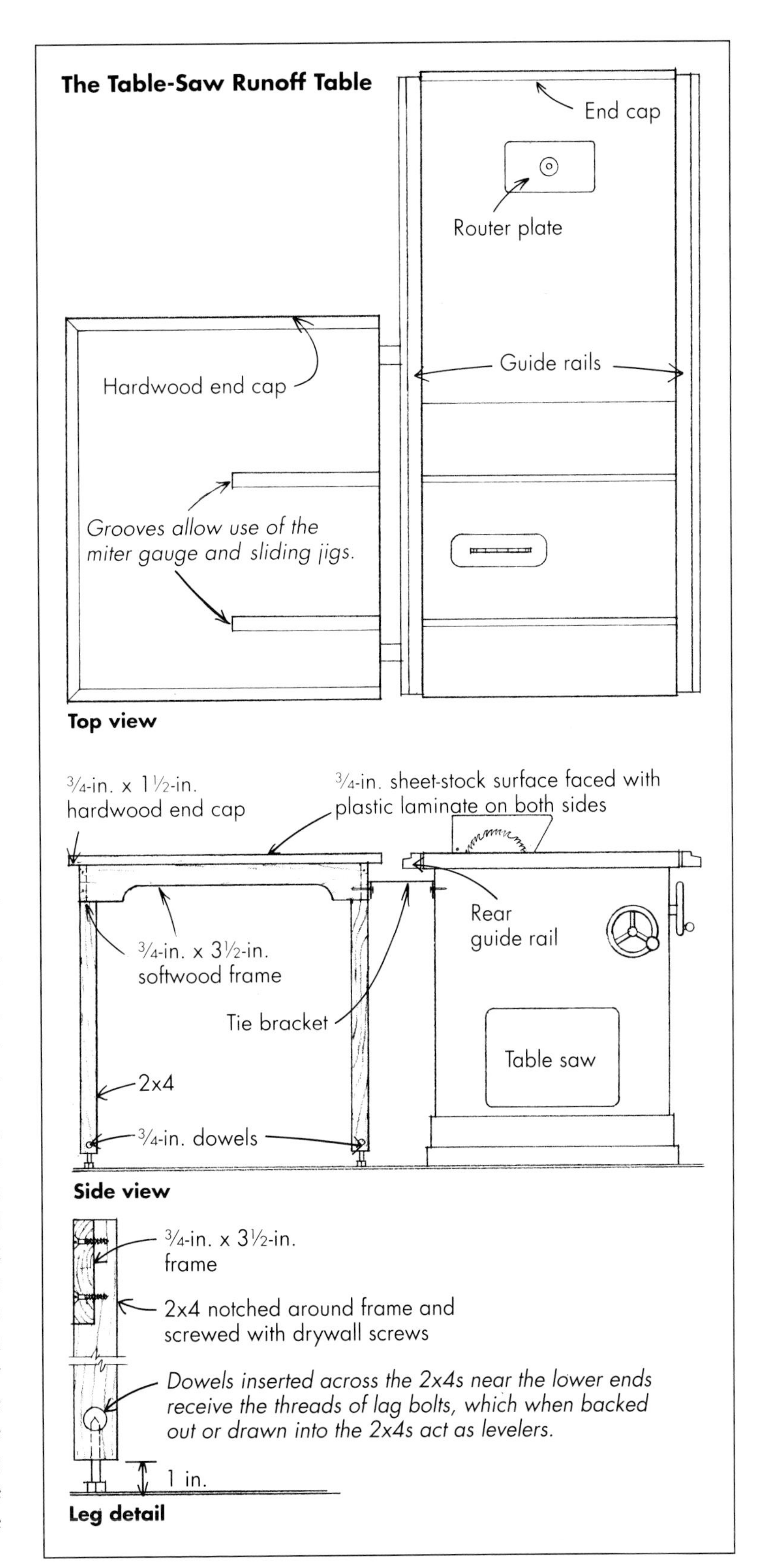

The runoff table stands independently and is tied to the table saw by screwing directly to the guide rail. If this isn't possible, bend two 1/8-in. steel straps to span the gap between the frame of the runoff table and the metal shroud of the saw. The grooves cut in the surface of the runoff table allow the unobstructed use of sliding jigs and the miter gauge.

The Radial-Arm-Saw Extension Table

The radial-arm saw's extension table (see the drawing on the facing page) is actually composed of three separate units tied together with a full-length fence. The fence acts as a backstop for the stock being crosscut and carries an indexed sliding stop that fixes the length of the material. The central unit supports the saw, while the side units keep the material level with the table saw and its attendant tables. Build the side units from 2x4s and make them as long as space permits—I recommend at least 8 ft. to the left of the blade and 6 ft. to the right. Unlike the table saw's runoff and extension tables, the surface of the radial-arm saw's table is not plastic-laminated, but is made instead of 3/4-in. sheet stock faced with 1/4-in. thick vinyl. This allows you to create a deep sawdust escapement between the table and the fence. It also enables you to replace the area that is constantly scored by the sawblade. (I simply stick down a 2-in. wide section of the 1/4-in. vinyl stock with double-adhesive carpet tape.)

Do not begin construction of this table until you've built and leveled the table saw's runoff and extension tables. Then extend a level line from the table-saw surface to the wall against which the radial-arm saw will rest. Measure down 1 in. from this line (to account for the 3/4-in. sheet-stock table with its 1/4-in. vinyl facing) and extend a level line along the full length of the wall.

Start construction with the central unit, which will support the metal frame of the saw. To find the height of this unit (composed of a 2x4 framework fastened to the wall at the back and supported in front by legs with adjustable feet), measure the height of the metal frame and subtract it from the second reference line. Mark a new level line on the wall and build the unit to this measurement. Next build the two side units to the height of the reference line drawn 1 in. below the table-saw reference line.

Secure the backing table in position and then construct the full-length fence. Build the 1½-in. wide by 2¼-in. high fence by laminating 3/4-in. stock, overlapping lengths and gluing and screwing the assembly together. (Remove the screws in the way of the sawblade after the glue has dried.) Build the fence on the table framework, which, having been leveled, should be perfectly flat.

Secure the fence to the backing table with 3-in. drywall screws, placing shims in between to keep the fence straight. (Any curvature in the fence will sabotage the 90° crosscuts you want to produce.) Place an untapered 1/4-in. shim near either side of the blade, and then work your way out toward each end of the fence, adjusting tapered shims in and out on 16-in. centers. A finish nail driven through the fence next to each shim will hold the fence in place temporarily yet allow some adjustment. Leave the head protruding at least 1/4 in. Continue adjusting the shims until a long straightedge or taut string indicates that the line is good and straight. Then sink the finish nails and screw through the fence and shims into the backing table. The 3/4-in. sheet-stock front table can then be laid in place on the support framing and screwed in place.

Next install the right-to-left-reading stick-on tape (available from Biesemeyer with their sliding stop, or as a single item through Woodworker's Supply; see Appendix I on pp. 140-142). Measure out 12 in. from the sawblade to the left along the face of the fence, make a mark and then square it up and over the fence with a combination square. Use the re-

The Radial-Arm-Saw Extension Table

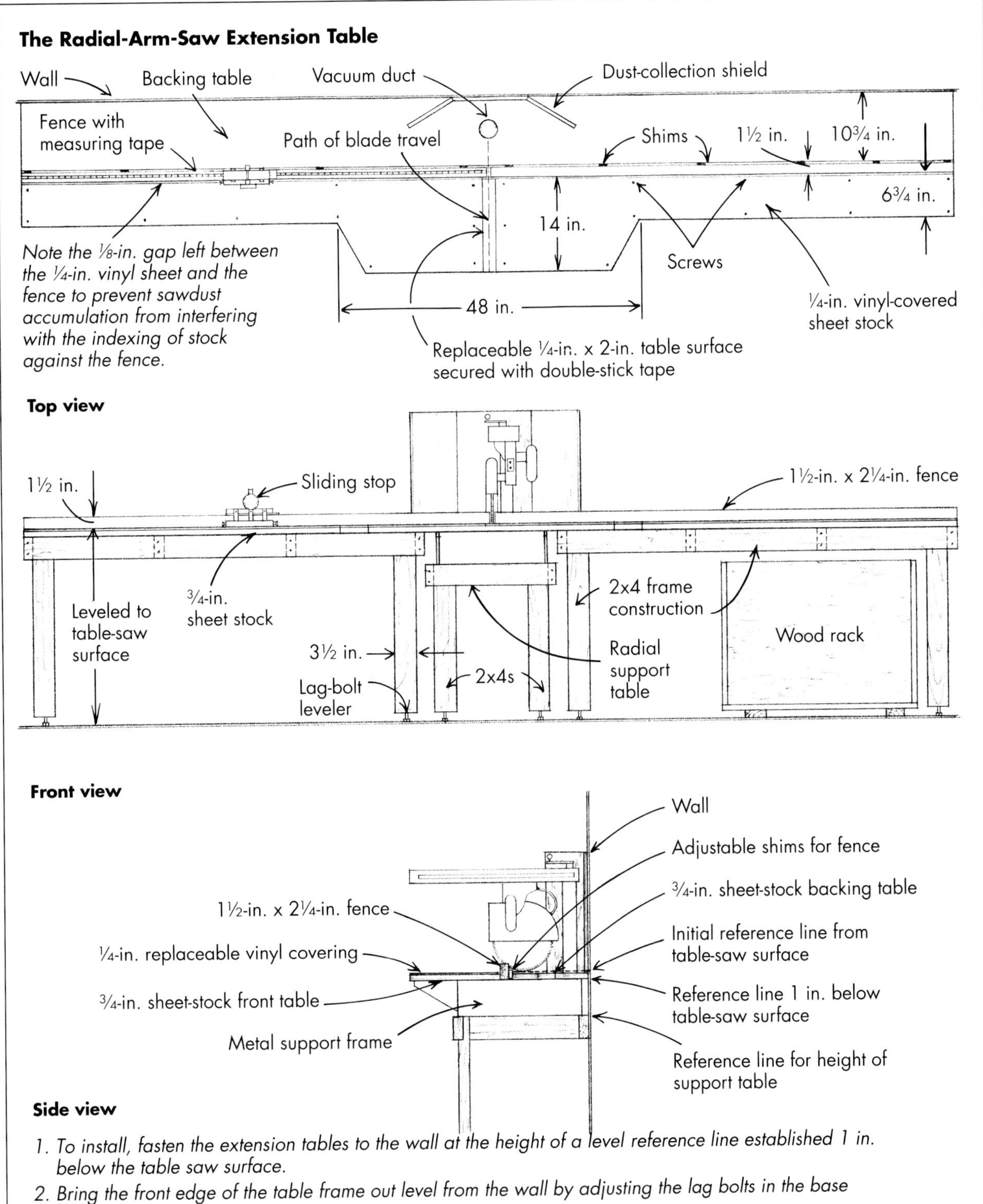

1. *To install, fasten the extension tables to the wall at the height of a level reference line established 1 in. below the table saw surface.*
2. *Bring the front edge of the table frame out level from the wall by adjusting the lag bolts in the base of the 2x4 leg supports.*
3. *Screw the ¾-in. sheet-stock backing table in place and fasten the full-length fence to it through the shims.*
4. *Install the front table, overlay the ¼-in. vinyl-coated sheet and screw it in place.*

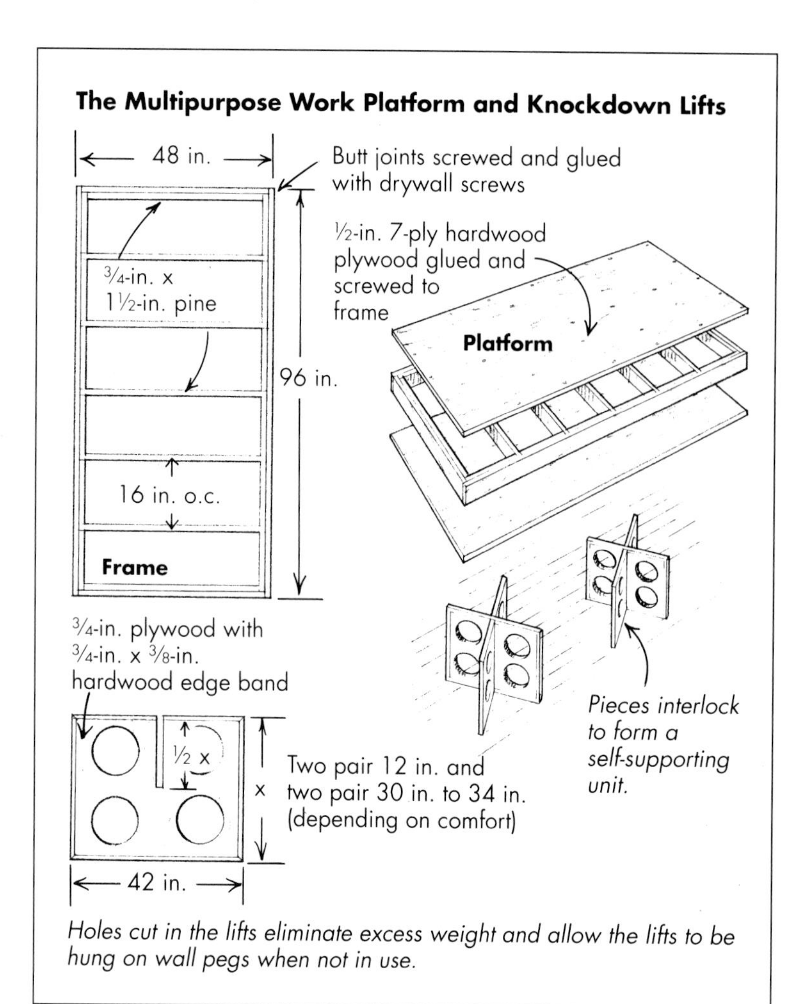

Holes cut in the lifts eliminate excess weight and allow the lifts to be hung on wall pegs when not in use.

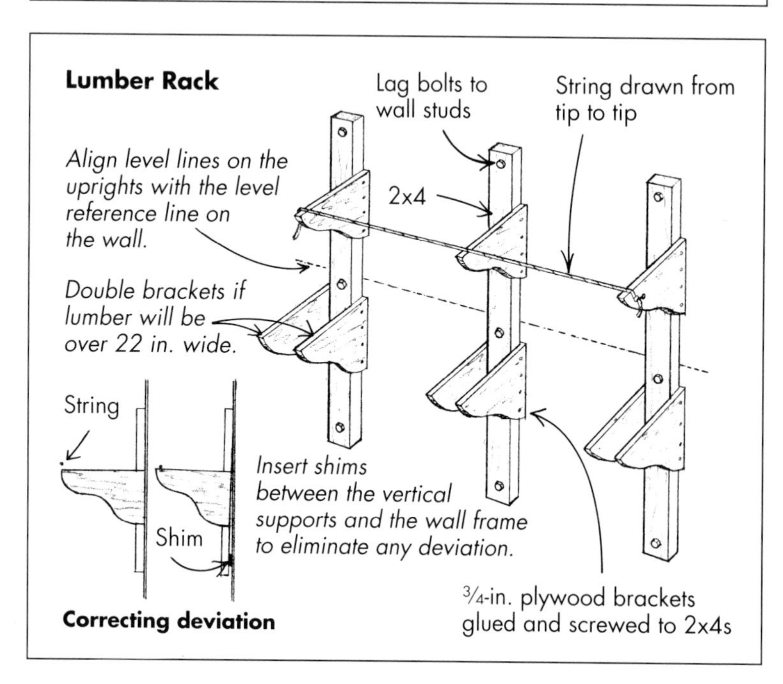

sulting line to index the 12-in. mark on the tape. Carefully begin sticking the tape down, rolling it out first to the right and then all the way to the left. Don't worry if you come out a little off the mark: The sliding stop has ample adjustment to allow fine-tuning of the setting during test cuts.

The Multipurpose Work Platform

The 4x8 work platform serves as a feed platform for sheet stock when it is adjacent to the table saw (as shown in the photos on p. 13). When relocated and raised, it acts as a component-assembly table; relowered, it becomes a carcase-assembly platform. Pairs of simple knockdown lifts hold the platform at the two heights.

The platform is built like a massive, hollow-core door (see the top drawing at left). It consists of a frame of ¾-in. by 1½-in. softwood pieces sandwiched between sheets of plywood glued and screwed to each side. Void-free, 7-ply ½-in. hardwood plywood is ideal for the top surface; $\frac{3}{16}$-in. plywood is adequate for the bottom. The result is a relatively lightweight but immensely strong, warp-resistant surface.

I borrowed the idea for the knockdown lifts from laminate installers, who often deal with large sheets of this unwieldy material. Construct the lifts from almost any grade of ¾-in. plywood. Go the extra mile and apply hardwood edging to the perimeter to reduce splinters and add durability.

The lifts also work singly (without the platform), to hold various components during processing and finishing operations. You can build more lifts at a variety of heights to serve additional support functions as the need arises. Color-code the pairs to avoid cross-matching.

Storage Systems

It's important to have storage racks and cabinets in the workshop to encourage you to put tools and materials away.

Lumber rack—Make a lumber rack from 2x4 uprights and some simple ¾-in. thick plywood brackets. If the lumber will be 22 in. wide or more, double the brackets to support the extra weight. Mark each upright in the same position; when installing, align these marks with a level chalkline snapped along the wall. Lag-bolt the uprights to the wall studs, installing the end uprights first: Make sure to plumb the brackets out from the wall as shown in the bottom drawing on the facing page. Plumb the intermediate uprights by aligning their tips with a string pulled taut from the tip of one end bracket to the other.

Clamp rack—Make a clamp rack (top drawing at right) from 2x4s and a closet pole. Build the rack as wide as necessary to contain your collection of pipe and bar clamps. The closet pole will hold as many C-clamps as you are ever likely to own. Lag-bolt the rack to the wall studs through the nailers notched into the uprights.

Sheet-stock bins—This rack (bottom drawing at right) holds sheet stock between 2x3 grids nailed to the floor and ceiling of the shop. I install ½-in. thick plywood strips to help slide the panels in and out. Build the rack as tall as the ceiling allows.

Storage cabinets—The cabinet below the sharpening bench (see the top drawing on the next page) might be the only full-scale example of your work on display when a prospective client walks into the shop, so do a decent job on it. I find that cabinets chock-full of drawers are far more useful than those that simply provide an enclosed storage space. Some of the drawers can be made quite large, allowing storage for an assortment

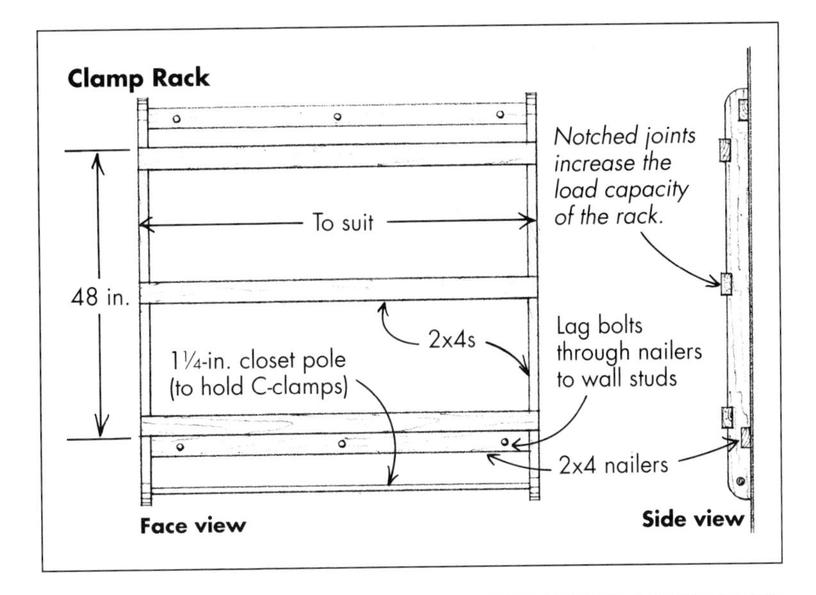

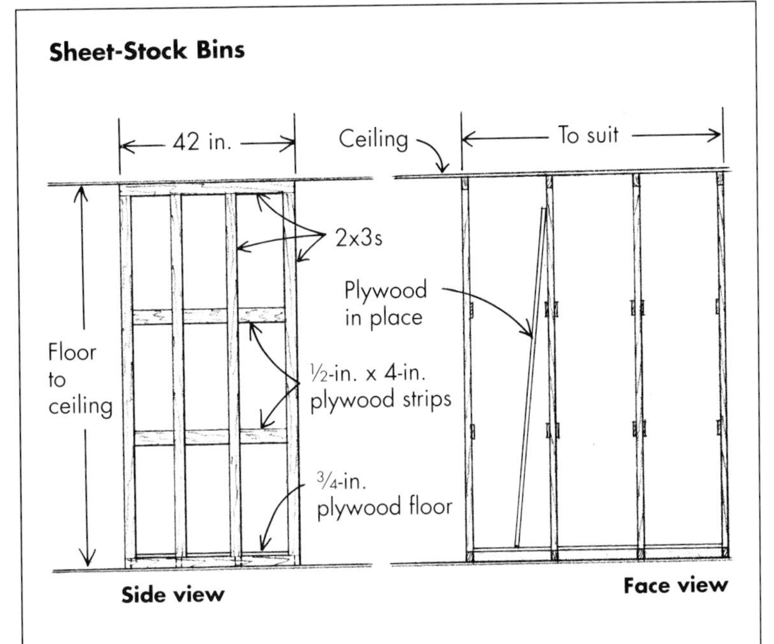

The plywood strips on the sides of the framing and the ¾-in. stock on the floor frame help ease and guide material in and out of the bins.

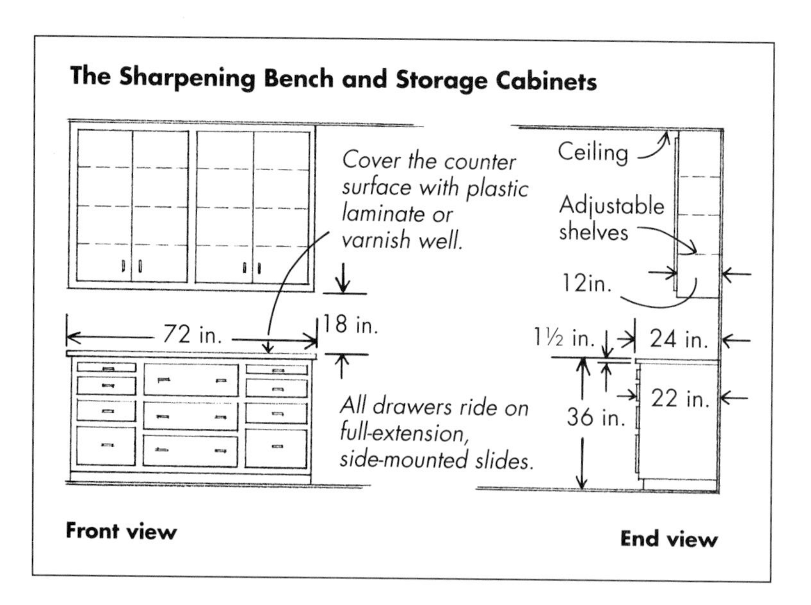

The Sharpening Bench and Storage Cabinets

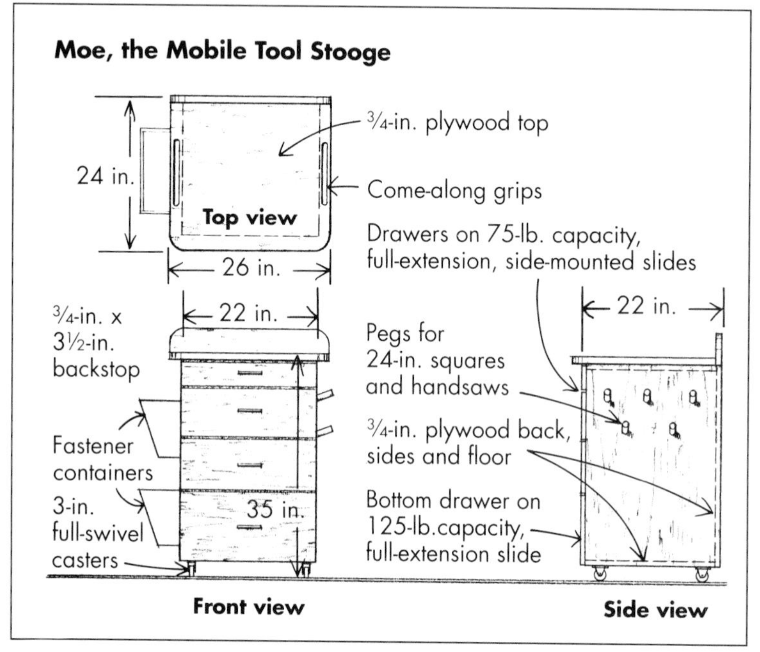

Moe, the Mobile Tool Stooge

of hand-held power tools and the opportunity to show off slick, heavy-duty, full-extension drawer slides. Since Murphy's Law states that the object of desire will always be found in the back corner of any drawer, it's not a bad idea to incorporate full-extension slides in the other drawers as well. (You can cheat by using a ¾-in. extension slide and building a shallower drawer.) Finish the cabinet with a water-resistant finish such as lacquer or polyurethane varnish, because the presence of the waterstones above will ensure a steady trickle of liquid onto the cabinet face below.

The cabinet above the bench can be your standard-issue, adjustable-shelf "upper." Build it right up to the ceiling and fit it with a good stock of shelves (you will eventually use every inch of them). Make a nice set of doors for this cabinet to keep out dust and to give you something pleasant to look at.

The Three Stooges

Wherever you go and whatever you do in the shop, there's a good chance that Moe, your mobile tool stooge, will be right there by your side. Like the sharpening bench's lower cabinet, Moe will be of most use if all storage is provided by drawers riding on full-extension slides. Build multiple divisions within the drawers to contain loose items.

Construct this stooge stoutly of ¾-in. sheet stock and give it four good-quality, full-swivel 3-in. diameter wheels (see the bottom drawing at left). Cutouts in the top overhangs provide convenient hand grips. Bins contain an assortment of fasteners; pegs hold handsaws and squares. Careful workmanship will allow the cart to withstand years of use, and a durable finish will keep Moe looking good.

Larry, the materials stooge, is the muscle of the trio. Larry will carry an incredible amount of raw materials and components from one production station to the next. To do the job well, this stooge must be carefully and sturdily built. Since Larry's structure is an open framework, obtain dimensional stability by using relatively large, tight-fitting and thoroughly secured joinery, as shown in the top drawing on the facing page. Construct the upper platform and the lower shelf frames from ¾-in. by 4-in. pine boards; join them with glued and screwed spline biscuits. Through-bolt the completed frames to the L-shaped legs and then cover them with ½-in. ply-

A family portrait of the three stooges: from left to right, Larry, Moe and Curley.

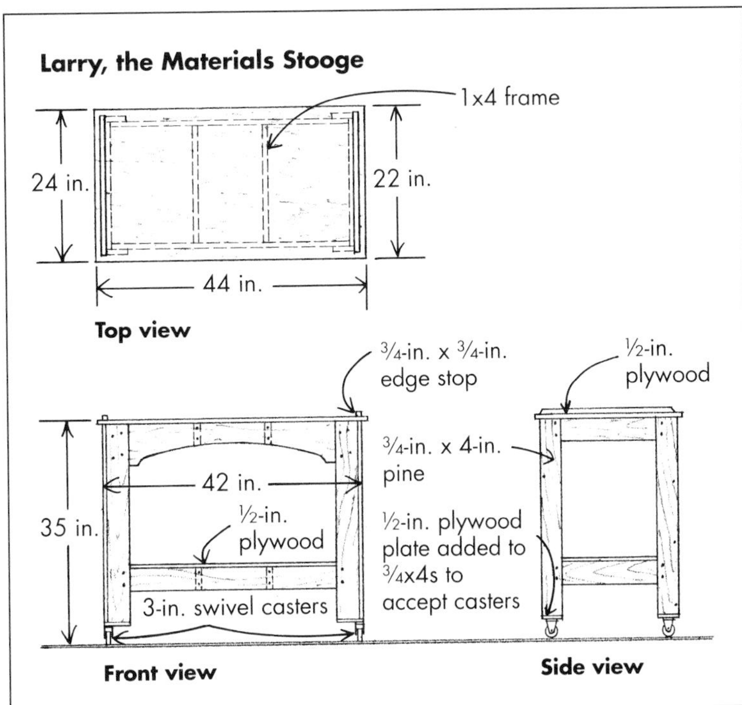

wood. The small strips at each end of the top platform keep material in place as Larry grooves from one scene to the next. Since grooving is groovier with larger wheels, especially for heavier fellows, provide this stooge with 3-in. full-swivel casters.

Last is Curley, the mobile trash bin. Curley is constructed from ¾-in. plywood fastened around a 2x4 base frame riding on four 2½-in. full-swivel casters (see the bottom drawing at right). Because this stooge spends a lot of time hiding under the table saw's extension table, you can construct it from scrap plywood or surplus planking. Screw the structure together, using support blocking in the corners. Note that the height of the bin is determined by the distance between the floor and the plunge router hanging below the table.

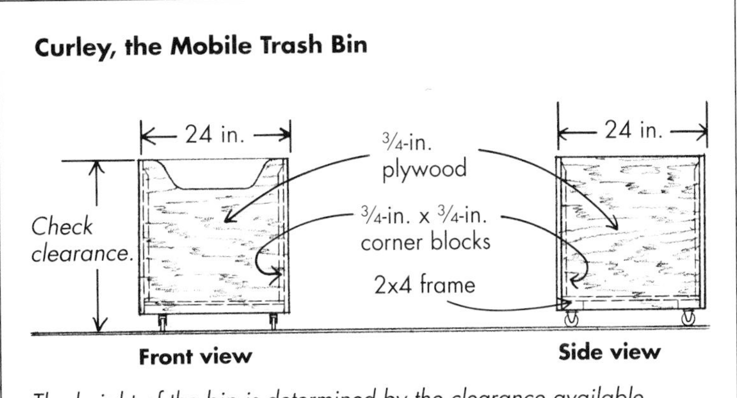

The height of the bin is determined by the clearance available between the shop floor and the underhanging plunge router.

Chapter 5: Safety in the Workshop

There's little chance you'd be doing this kind of work unless you really enjoyed it: The hours are long, the tools are loud, the air isn't fresh, there's no boss to bellyache about ... but you love every minute of it. Perhaps too much so—I rarely feel there are enough hours in the workday to accomplish everything I'd like to. There's nothing wrong with this energetic approach, except when good intentions surpass your physical ability to keep up. There are those times when the body is simply too tired, too mesmerized by the drone of the tools or too lacking in good, fresh air to work safely any longer. The situation can be exacerbated by illness, emotional distress or the ingestion of anything that overstimulates or depresses the nervous system.

The tools cabinetmakers work with can be dangerous. To be operated safely, they need the attention of an aware mind and a healthy body. If you are not up to par, you must stop working. The risk is simply too great.

Protective Clothing and Devices

Even on the best days, we need to protect ourselves from our machines and their by-products. The first thing I do when I enter the shop is put on a work apron that covers most of my torso, to keep splinters and sawdust off my clothes. I recommend the apron made by Bridge City Tools (see Appendix I on pp. 140-142). It's durable and has plenty of pockets, and the design of the straps takes the weight of the apron's stash of tools and fasteners off your shoulders and puts it more comfortably around your waist.

Your feet also deserve consideration. Sneakers are comfortable, but they offer no protection from objects that may drop on your toes. When I'm in the shop, I wear lightweight, steel-toed workshoes with a padded arch-support insert for cushioning.

Have you noticed how loud your tools are as they scream their way through the wood? You shouldn't, and not because you're already deaf, but because you're wearing some form of ear protection. I use foam earplugs, which are effective and comfortable. (I find earmuffs uncomfortable over the sidepieces of my glasses, and they make me feel claustrophobic.) Earplugs are inexpensive and can even be washed and reused. For me, the type made from supple material, with rounded ends, is best. (Mine are made by Moldex-Metric Inc.; See Appendix I on pp. 140-142.)

When you work with fast-moving, debris-spewing machinery, you must protect your eyes. Since I wear glasses anyway, I use a safety glass cut to my prescription and mounted in a sturdy frame. Side shields keep debris from sneaking in the sides. If you don't normally wear glasses, do so when you are in the shop. Goggles scratch quickly and are uncomfortable to wear for any length of time; you simply won't get into the habit of wearing them throughout the day.

You must also protect your lungs. The air in cabinet shops can be toxic if you are allergic to certain wood species, glue substrates or finishing materials. The dust-collection system should eliminate a good deal of the dust-related toxins, but when dust concentrations in the air become visible you should be wearing a mask. The best mask I've ever encountered is #8560 made by 3M for non-chemical-mist protection (see Appendix I on pp. 140-142). It features two separate flexible straps, an extra-thickness shield membrane and a foam cushion that form-fits over the bridge of the nose. It's essentially a deluxe version of those useless single-strap dust masks.

If you are dealing with toxic fumes, a dust mask is ineffective, and you need to wear a rubber-shield, carbon-filter respirator. These aren't very comfortable, but they do work. Change the filters regularly if you are exposed to high concentrations of fumes. Of course, the best protection comes from eliminating hazardous materials from the shop. In particular, I avoid formaldehyde-laden particleboard. I've also stopped using solvent-based finishing products in favor of those that are water-based .

To protect against fire, keep at least two fully charged ABC-type fire extinguishers (good for all types of fires) in separate, readily accessible locations. Don't keep any near the stove—because this is a likely source of a conflagration, you may not be able to get to them when you need them most.

First Aid

A well-stocked first-aid kit in an unobstructed area in the main portion of the shop is an absolute necessity. Buy a kit that provides a dust-free storage container. In addition to the usual assortment of Band-Aids, sterile gauze pads and tape, include a tourniquet, scissors and a good pair of tweezers for splinter removal (I also carry tweezers in my apron pocket). Since you might be alone when an accident happens, you should make it a point to learn basic first aid. Your local Red Cross chapter will probably offer an inexpensive course; let the instructor know what you do for a living, and he or she may be willing to give you some additional, shop-specific instructions. If you cannot get to a class, study a good book on first aid (see Appendix I on pp. 140-142).

Because hand and eye injuries are common in woodworking shops, post the telephone numbers of doctors who specialize in these areas, along with the number of the closest emergency room, right next to the telephone.

Above all, be ever conscious of safety and be continually careful: An old boatbuilder colleague who still possesses all his fingers once told me that every morning before going to work he would make a point of looking at his hands and promising "to bring 'em all home that night."

Place a dustproof first-aid kit and an ABC-type fire extinguisher in an easily accessible area in your shop.

Section II

The various stages in Tolpin's production process are carefully sequenced to make the most efficient use of his time, machinery and modest floor space.

The Process

Once you've found a suitable space in a promising location and tuned up the tools and fixtures, it's time to pass out the cigars: A shop has been born. But before slapping the first switch and breathing some life into this baby, give some thought to developing a highly productive work style—the next step in the creation of a successful woodworking livelihood.

To make the shop really work for you, and to enjoy the benefits of an efficient and sensibly paced production process, the sequence of tasks involved in building a set of cabinets must be carefully organized. You must scrupulously delineate how materials will flow through the shop during the project: Products of one operation must not block access to the machinery and space requirements of the next operation; components, related or not, must be grouped so that a particular tool setup need be created only once; and operations must be organized so that all materials reach the same production phase at the same time.

The following chapters present the production process I currently use in my shop. After an introduction to the elements of successful cabinet design and the "block production method," each subsequent chapter represents a phase of the process. These phases, or production blocks, are graphically represented by flow charts, which illustrate the relationship of steps within a block. The production process in its entirety is consolidated in the master flow chart that appears in Appendix III on pp. 144-145.

Chapter 6: Cabinet Design and Block Production

If you are not familiar with European-style cabinets, I suggest you find a showroom that has them on display and take a good look. The first thing you might notice is that the spacing between doors, drawer faces and fixed panels is extremely close, usually less than ⅛ in. When you open the doors and drawers, you'll see the hardware that makes these close tolerances possible. Hinges are adjustable in three directions, drawer slides adjust up and down and even the fasteners that attach the drawer faces to their boxes allow adjustment. A screwdriver is all that is necessary to fine-tune the relationship of all the face components to each other.

Looking further, you'll find that the drawer slides and door-hinge-mounting plates, as well as shelf-support clips and connection fittings (used to hold panels in place or the cabinets together), all fit into identical holes spaced precisely 32mm apart in two vertical lines on the inner sides of each cabinet (see the bottom photo on the facing page). These holes, which measure 5mm in diameter, permit the cabinet sides to receive a variety of fittings, allowing any one component to hold many possible configurations of doors, drawers and adjustable shelves. Because the Europeans standardized their hardware systems (a result of the need for massive reconstruction of housing following World War II), they were also able to standardize cabinet components. The implications of such standardization for the mass production of cabinets are obvious.

While the technologists were developing these new hardware systems, the cabinetmakers were busy eliminating many traditional design features. Face frames and certain moldings were deemed unnecessary, as were integral kickboards and support frames (kickboards could be replaced with a facing board attached to adjustable leg hardware with clips). The final product was highly functional, straightforward and incredibly efficient to build.

It was also a pretty stark piece of work (some would even say sterile). The monolithic face presented by a run of European-style cabinets, broken only by fine lines between the unadorned, flat drawer faces and door fronts, offers little to please or interest the eye. Although the typical Eurostyle cabinet can work well in a contemporary setting (and certainly in institutional applications), it can easily seem out of place in traditional-style houses.

In my adaptation of European construction methods, however, I've found no limit to the variety of styles that can be produced. As you can see in the photos on pages 39 and 40, the appearance of these kitchens, recent products of my shop, are decidedly traditional. Beaded panels, 19th-century-style cock beading and a variety of other molding elements abound, yet the cabinets' foundation is European contemporary.

This traditional-style set of cabinets is based on contemporary European construction and hardware systems.

A look inside a cabinet reveals the European cup hinges, adjustable shelf clips and drawer slides. All these fittings are located in two rows of 5mm holes drilled vertically in 32mm increments.

Cock beading, mid-rail and pilaster moldings between door and drawer faces add a traditional touch to European-style cabinets. (Photo by Richard Slack.)

I offer a number of other styles of cabinetwork to my clients as well. The majority of these are interpretations of traditional American designs, yet each is based on the same basic module using 32mm-system hardware. Each module is composed of two sides, a floor, a top frame (or sheet-stock ceiling if the cabinet is an upper unit) and a back, as shown in the top drawing on p. 41. Before the module is assembled, the sides are drilled with the vertical rows of 5mm holes spaced 32mm apart. The first row is set back 37mm from the front edge of the case to accommodate the cup-hinge mounting plates and the front screws of the drawer slides. The second row is set back far enough to catch the rear mounting hole of the drawer slides. (For example, 22-in. slides require a 517mm setback.) Because I use European adjustable legs, I also drill holes for these in the cabinet floors before assembly. And as I don't have room in my shop to finish assembled carcases (nor do I want to increase overhead by having that room), I prefinish the components of each module and install all hardware before assembly.

This basic module serves as the foundation for any style of cabinet I wish to produce. A number of moldings can be used to augment and enhance the design and to tie the modules together visually (see the bottom drawing on p. 41). The cornice, light pelmet (which also serves to hide under-cabinet lighting systems) and plinth offer strong horizontal lines that can bridge an entire run of cabinets, while pilasters, spacers and an applied mid-rail break up the spacing between face components.

In this type of cabinet, the exterior surfaces of the module sides are never seen: They are either hidden against a wall or another cabinet, or covered with an applied panel. Using applied panels allows exposed ends of cabinet runs to mimic the style of the doors or to present a beautiful surface of solid wood.

Devoting some attention to these design elements when developing your own line of cabinetry will enhance the uniqueness and visual appeal of your product. I suggest beginning with large-scale (or even full-scale) drawings, and then building prototypes of the more promising designs. (See Appendix II on p. 143 for an excellent book on furniture design.) Do a good job when putting together your prototypes: If you should go with the design, they will likely become product samples.

As you can see, the way I developed my cabinet designs was profoundly affected by the cabinetmakers of postwar Europe. The challenge remains in coming up with a production process that allows custom woodworkers to build these kinds of cabinets within the parameters of our humble spaces and modest tooling. Over the past ten years, through much thought and trial and error, and with some heaven-sent help from tool manufacturers (who have brought small-shop versions of high-tech industrial cabinetmaking tools within grasp), I have worked out my own solution to this challenge. I have even given it a name: the Block Production Method.

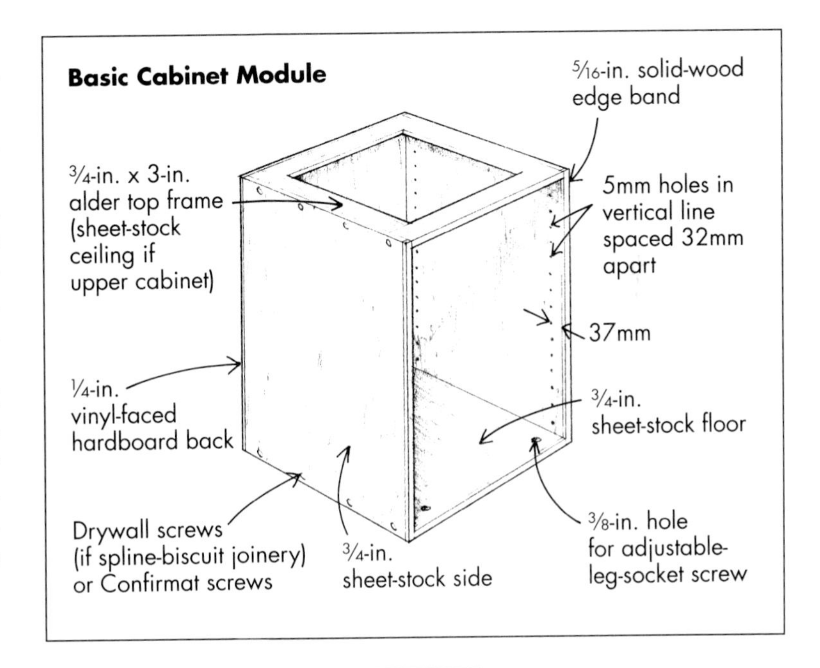

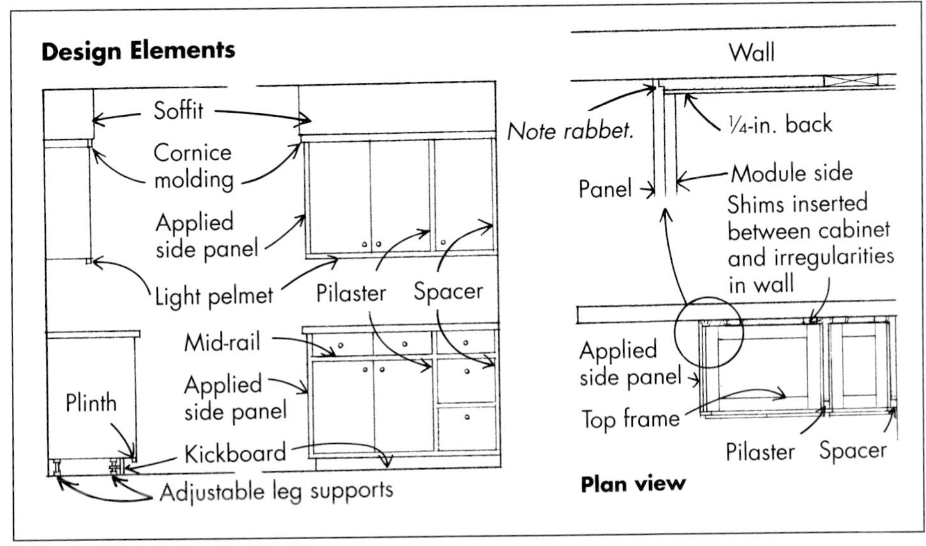

The Block Production Method

For a small-shop operator to cope with the immensely varied, space-consuming operations involved in producing a set of cabinets, it is necessary to develop some form of consistent production methodology. The amount of time spent handling materials must be minimized, and processing operations should flow in a smooth, nonrepetitive stream. Operations must be organized so that products of previous operations will not hinder access to work stations. And in meeting these goals, it's important not to lose control of the quality of the results.

My production method is based on a highly organized sequence of steps that details the processing of material from its entrance into the shop to its exit as a part of a finished cabinet unit. The premise of this methodology is that all the tasks involved in the manufacture of a set of cabinets can be grouped into exclusive process "blocks." It ruthlessly demands that all the materials intended to undergo a certain operation be grouped together, even if they are totally unrelated. Thus shelf edgings may find themselves in company with drawer faces and door stiles in a crosscutting operation. Only one tool setup and subsequent space commitment is delegated to accomplish a major element of the production process. The blocks are carefully sequenced so that the material flows smoothly throughout the work space, and also so that subsequent steps occur in the proper order.

The flow chart on the facing page shows the sequence and interrelationship of the production blocks. Of course, this representation is generalized, as each block shown on the chart actually contains its own, detailed flow chart. (See the flow charts that accompany the following chapters, and the master chart in Appendix III.)

In addition to the vast time savings that have resulted from implementing the block production method, other benefits have also arisen. Because the processes are so clearly delineated within the context of any one project, it becomes possible to accurately generate data on the amount of labor required in any given process. (This is valuable information when estimating the costs of proposed projects, as discussed on pp. 121-122.) Another benefit is that breathing spaces are automatically built into the pace of the work. Between blocks, you can catch up and clean up; in the humble spaces of most independent cabinetmakers, this really makes a difference in the quality of work life. Finally, once the block production method is firmly established in the shop, it becomes possible to do more than one project at a time without confusion. A small job automatically nets a higher profit margin as it rides through the tool setups on the coattails of a larger project.

In the ensuing chapters of this section, you'll find that the elements of the block production method are presented in the sequence they occur in the course of manufacturing a set of cabinets in a one-person shop. As you read through, periodically refer to the master flow chart on pp. 144-145 to help you maintain a clear perspective of how each step relates to the rest of production. I keep a large copy of the flow chart hanging on the wall of my shop, lest I forget where to turn at the end of each production block.

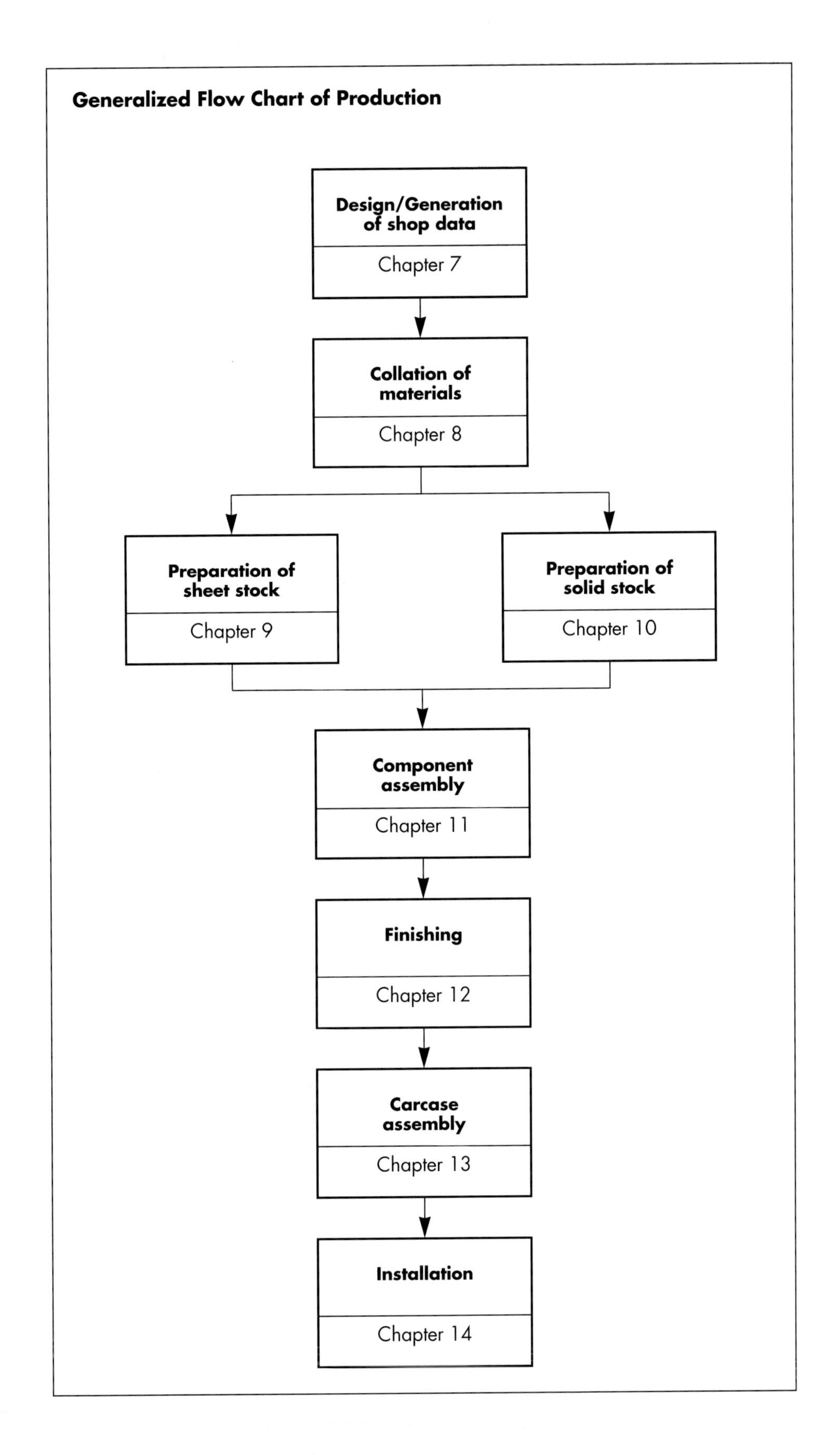

Generalized Flow Chart of Production
Design/Generation of shop data
Chapter 7
Collation of materials
Chapter 8
Preparation of sheet stock
Chapter 9
Preparation of solid stock
Chapter 10
Component assembly
Chapter 11
Finishing
Chapter 12
Carcase assembly
Chapter 13
Installation
Chapter 14

Chapter 7: Generating Shop Data

The beginning of any project invariably involves at least one client consultation. The purpose of the consultation is to determine the exact specifications of the product you are being hired to build. (For the benefit of all parties, this information should be in writing as discussed on p. 111.) During the consultation, you'll show the client photos of your products and samples of your standard door and drawer-face styles. On the rough elevation sketches, note unusual cabinet configurations, specialty hardware (such as lazy Susans and slide-out baskets) and the client's choice of hardwoods and decorative hardware.

If the client should ask you to design a new style of cabinet for the project, be aware that prototype development takes time. It may also require the purchase of additional equipment (such as router bits) and the construction of specialized jigs. You may absorb these costs or pass them on to the client. Unless I am certain I will add that particular style to my standard line, I charge the client for development time but absorb the cost of the tools, which will likely find other applications in the shop.

Assuming that the client hasn't already had a floor plan drawn up, and that the project involves more than one freestanding unit, the next step in design is to lay out the project in plan view. You can go one better and use the Stanley Tool Company's "Project Planner," available at many builder-supply yards, to present the client with a perspective view of the layout. (You might need extra pieces from a second kit to create a U-shaped kitchen.) Once you and the client agree on the layout, photocopy it and transfer any notations on cabinet configuration or specialty hardware to the copy from the rough elevation sketches. Your contract with the client will specify the dimensions of each cabinet, any specialty hardware and the choice of style, wood and finish.

Shop Drawings

Once you have received the go-ahead and a deposit (see p. 111), develop a scaled plan view of the project on vellum paper so that blueprints can be made. This view, which looks directly down on the cabinets, details the relationship of the units and allows you to delineate the modules within each cabinet run. Even if you're given a set of floor plans, it's safest to measure the space the cabinets must fit into yourself. If wallboard, final flooring or ceilings have not been installed, find out how they will affect your measurements. Also note anything else that might influence the design or installation of the cabinetry. Mark on your floor plan the exact placement of doors and windows if they will abut the cabinets; locate any electrical or plumbing fixtures; and note steps in floors or jogs in walls in the path of the cabinets. In addition, draw in the appliances, adding their model numbers and all pertinent dimensions. (Using the model numbers, I double-check with the manufacturer the dimensions of any appliances not already on hand.)

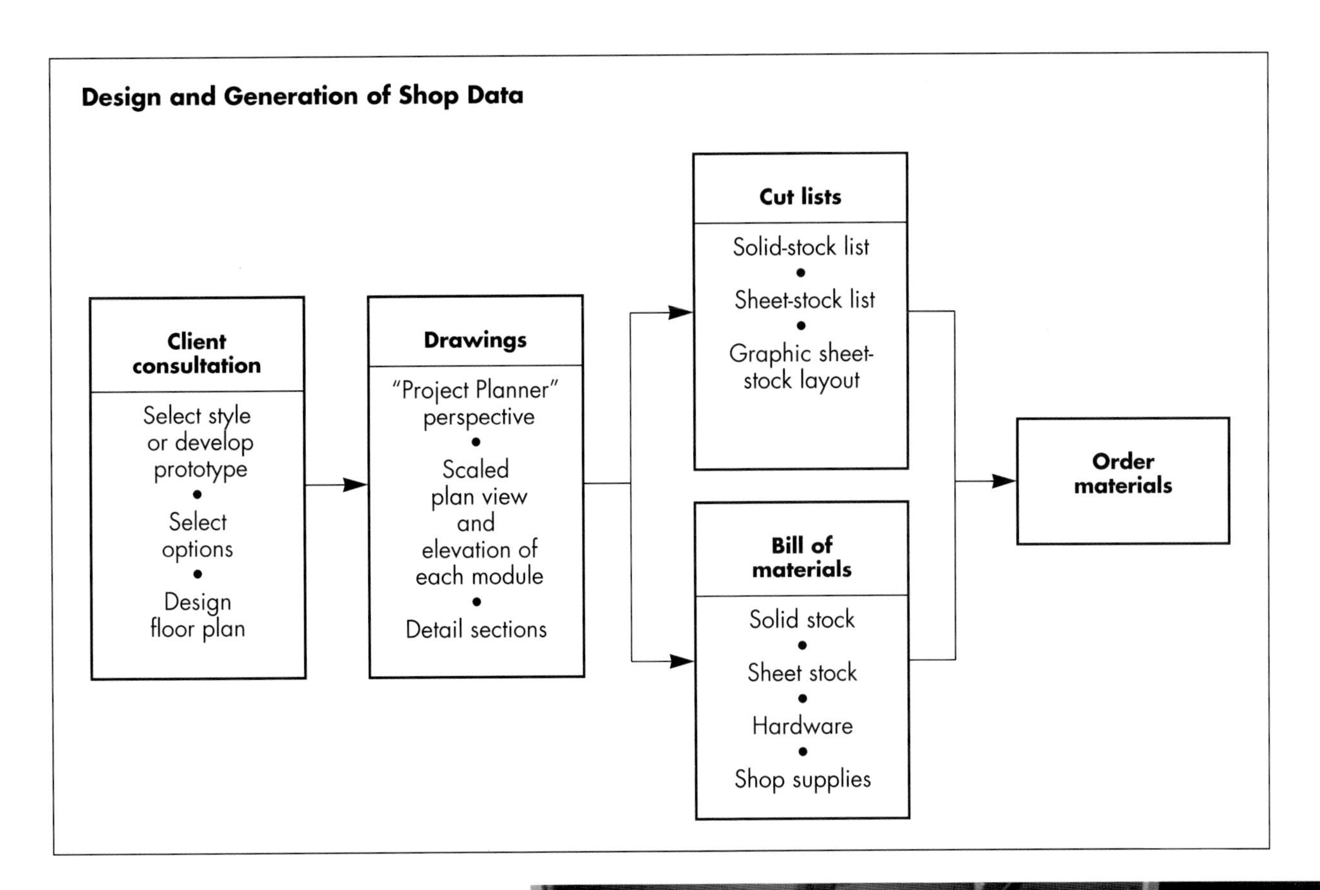

After defining the modules, assign each an identification symbol so they can be located easily when referred to in other shop drawings. I use the first letter of the client's last name followed by a number. Thus a kitchen built for Mrs. Smith that is composed of 12 modules will have designations of S1 to S12. Hang up a blueprint of the drawing somewhere in the shop for reference throughout the production process.

An appealing perspective view of almost any kitchen layout can be created easily with the Stanley 'Project Planner.' The result can be photocopied and given to the client.

Typical 5x8 Elevation Card

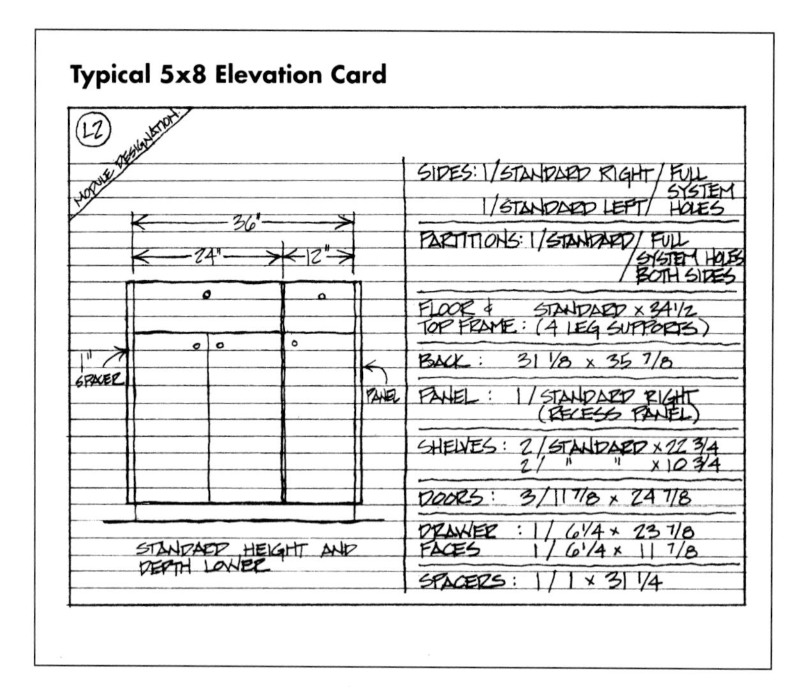

Typical Master Cut List for Solid Stock

Face-frame rails and stiles (oak)				Door (oak)	Edge bands (oak)	Exposed nailers (oak)	Top frame (alder)
3	2 1/4	1 3/4	7/8	2	5/16 x 13/16	3/4	2

Moldings			Drawer faces (oak)				Slide-out supports (oak)		
3/4 cove	3/4 scotia	2 1/4 light mold	9 1/4	6 3/4	5 3/4	4 1/4	2	1 3/4	1 1/4

Another necessary shop drawing is an elevation of each module to be built. (Since I've begun using the "Project Planner," I don't bother to draw elevations of the cabinet runs.) I do this on an elevation card, a 5x8 index card, which provides space for the drawing and a list of components and their dimensions (see the top drawing at left). I keep these cards handy in a file-card box on my friend Moe, the mobile tool cart.

Sometimes it is helpful to prepare elevations and cross sections of unusual details, such as built-up moldings. If you have not already done so, draw a side view of your standard cabinet component, including the spacing dimensions of the 32mm-system holes. Hang copies of these plans near the plan view of the job in progress.

Cut Lists

I develop two cut lists: one for solid stock (shown in the bottom drawing at left) and one for sheet stock. The information is compiled from the list of components on each elevation card. I organize sheet-stock components according to the type of sheet from which they'll be cut—thus, every panel to be cut from ¾-in. birch plywood will be listed under that category, and then arranged on a graphic representation of the 4x8 sheets, as shown in the drawing on the facing page. This step allows me to organize the panels to minimize waste, and provides a fail-safe, visual reference in the shop during actual panel sizing. Be sure to note the module designation and component function on each part.

When possible, lay out the panels so the initial cuts rip the sheets along their full length. This allows you to handle smaller sections of the sheets in subsequent sizing operations, ensuring a higher level of accuracy in the results.

Bill of Materials

To produce an order sheet for materials, compile the numbers from the cut lists. First, total the board footage of solid stock needed, using the solid-stock list and a calculator with memory capability. By entering in all the dimensions of the stock in feet (or fractions thereof), you can produce a sum representing the board footage.

For example, if a column of 1-in. by 4-in. stock adds up to 135 lineal feet of material, this would be entered into the calculator's memory as 1 x .333 (4/12) x 135. (The 4/12, or .333, accounts for the fact that a 4-in. wide board represents one-third of a board foot of material for each running foot of its length.) A column of 1-in. by 6-in. stock by 150 lineal feet is entered as 1 x .5 x 150. If the components will be derived from thicknesses of stock other than 1 in., this must be accounted for. For example, if 4-in. moldings will be cut from 1½-in. stock, this should be entered as 1.5 x .333 x the lineal footage. When the calculator is asked to sum its memory, the figure will be the total board footage of the material. When ordering solid stock, however, you should always add at least 15% to account for defects and wastage during processing.

Typical Graphic Sheet-Stock Layout for ¾-in. Case Components

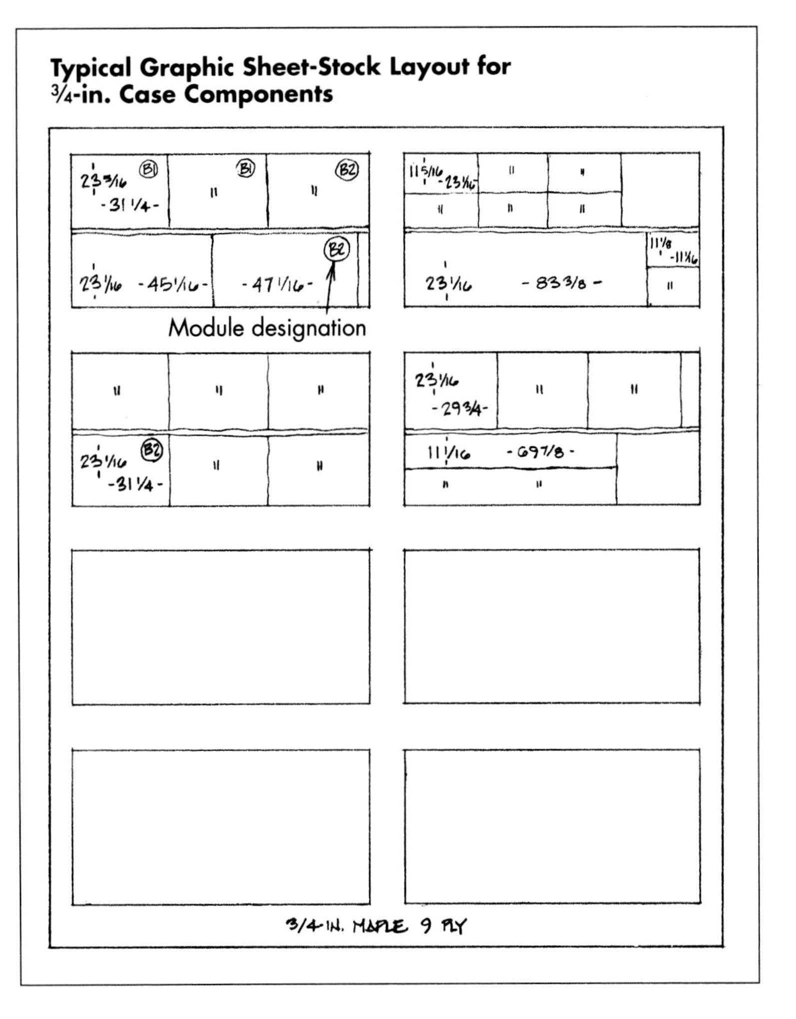

To figure out the sheet-stock order, count the types and quantities needed directly from the graphic representation. Add at least one more sheet of each type for insurance in the event of a processing error.

Determine the hardware required for a project from a tally of the elevation cards, which list the doors, drawers, shelves and specialty items present in each module. Estimate incidental hardware, such as connection fittings, cover caps and adjustable leg supports, in relation to the size and complexity of the module. Don't overlook the fact that a project will also consume shop supplies not listed on the elevation cards or cut lists. Estimate the amount of glue, fastenings, finish materials and applicators, sandpaper, steel wool and rags that will be necessary, and replenish any of these supplies when ordering the rest of the material.

To avoid delay during production, order all the material needed for a project as soon as possible. If you have taken a deposit while in the middle of another job, try to take a day off to make up the elevation cards, cut lists and bill of materials for the new project. Order the hardware immediately and arrange for your supplier to bring in the wood as soon as you have cleared the current project out of the shop.

Chapter 8: Collating and Stacking Materials

The presence of lumber racks and the 4x8 multipurpose work table makes it possible to avoid the small-shop gridlock that threatens at the beginning of each new project. It also helps to prepare the shop for the arrival of materials by purging the racks and floor space of surplus stock and scrap. It's my rule that solid stock measuring less than 2 in. by 2 ft. and any plywood offcuts less than 2 ft. square go either into the woodstove or into bundles to be donated to local hobbyists.

Clear the lower two levels of the lumber rack above the radial-arm saw in preparation for the arrival of the solid stock. When stacking the new lumber, separate it by thickness, placing the bulk of it (usually the 1-in. stock) on the first level; reserve the level above it for stock of other thicknesses. To prepare for the sheet stock, place the 4x8 platform on the 12-in. supports and orient it west to east, as shown in the shop layout plan on p. 9. Except for the material that will later be surfaced with plastic laminate, which is laid on edge against the south wall of the shop, all the incoming sheet stock will rest here.

To keep production flowing smoothly, the sheet stock that is stacked on the 4x8 platform must be properly organized. The general rule in collating is that the material used last in subsequent processing should appear at the top of the stacking order. Because you want the ¾-in. thick sheets for the carcase components, including shelves and slide-outs, to appear first for the next operation (predrilling system holes and other milling tasks), stack them on the bottom so they'll be sized last. After these sheets come the ½-in. thick drawer-side stock, ¼-in. birch (good one side) drawer bottoms, ¼-in. recessed panels and ¼-in. vinyl-faced back stock, in that order. Drawer fronts are usually of solid stock, but if they are plywood they are sized with the ¾-in. carcase-component stock. Any sheets that require plastic lamination are put on the pile after they have been laminated; consequently, they will be sized first.

Check the hardware as soon as it arrives, counting the pieces against the bill of materials for the project. If there is a discrepancy, check the packing slip to see if any items are on back order or if there was an error in fulfillment. Resolve deficiencies immediately to avoid delays when the production flow reaches the assembly blocks.

Except for large specialty items, store the hardware in the drawers of the sharpening-station cabinets, which are located near the area used for carcase assembly (see the layout plan on p. 9). Commit one large drawer to drawer slides and divide its length into four bins. The two bins on the left hold the left-handed portions of your most common slides (22 in. for standard kitchen lowers and 20 in. for vanity cabinets); the two bins on the right hold the right-handed parts. Leave other lengths of drawer slides in their shipping cartons and store them in a separate drawer. Sacrifice another drawer to cup hinges

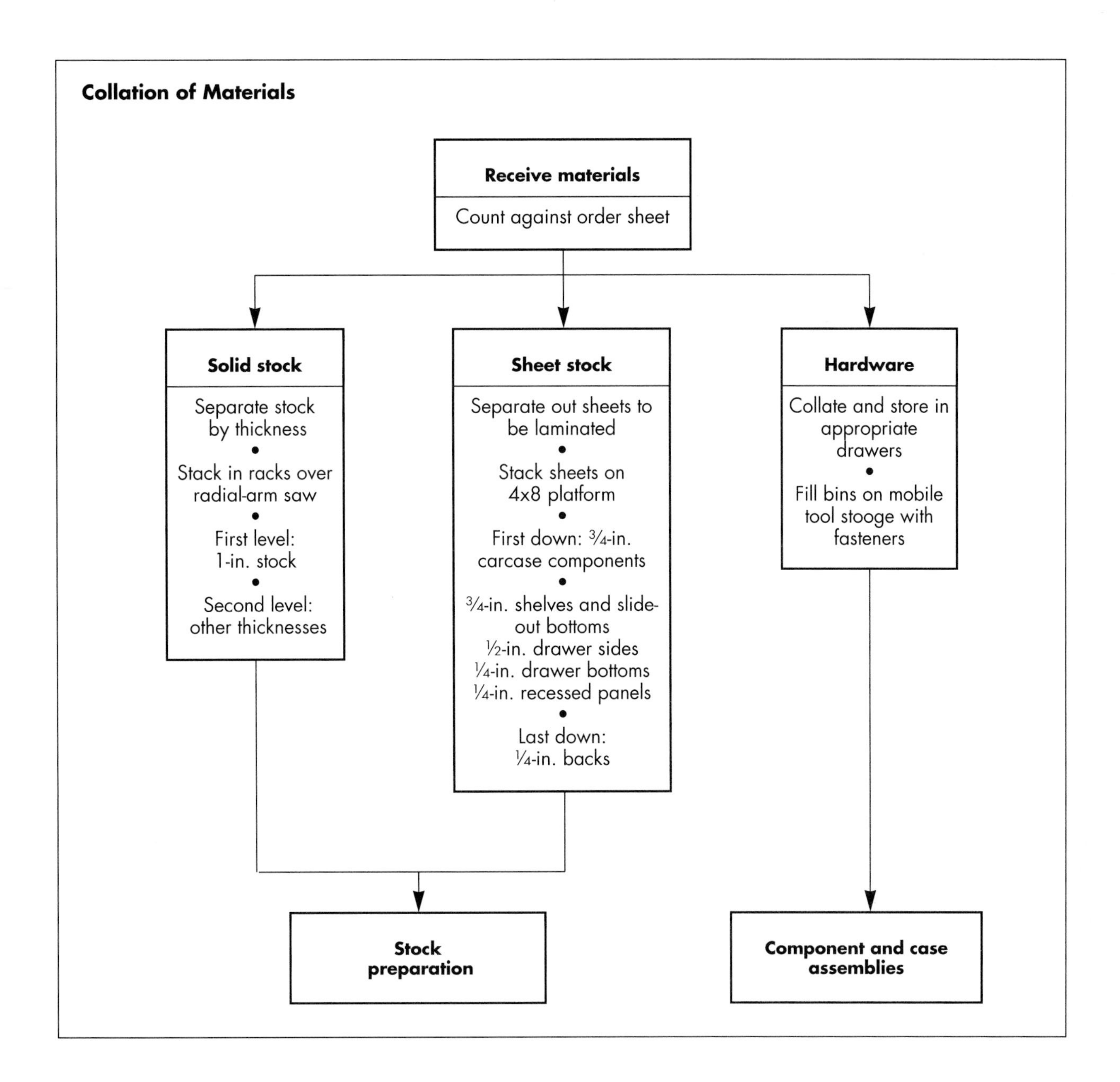

and their mounting plates. Store additional hardware in the shipping boxes. Put the adjustable legs and their sockets in the last drawer committed to hardware inventory, but keep the kickboard clips and cover caps separate. Stored with them, in separate boxes, are drawer-front adjusters, shelf supports, drawer bumpers, spacers and mounting sockets, bumper pads and an assortment of installation fittings and cover caps.

The fastening hardware used in component and case assemblies is stored in bins provided for this purpose on the side of Moe, the mobile tool stooge (see p. 32). Stockpile surplus fasteners in another drawer, if you have one to spare, or put them in stackable plastic bins and store in the upper cabinet above the sharpening bench. And that's it—you're ready to jump into the production flow and do some cabinetmaking.

Chapter 9: Preparing Sheet Stock

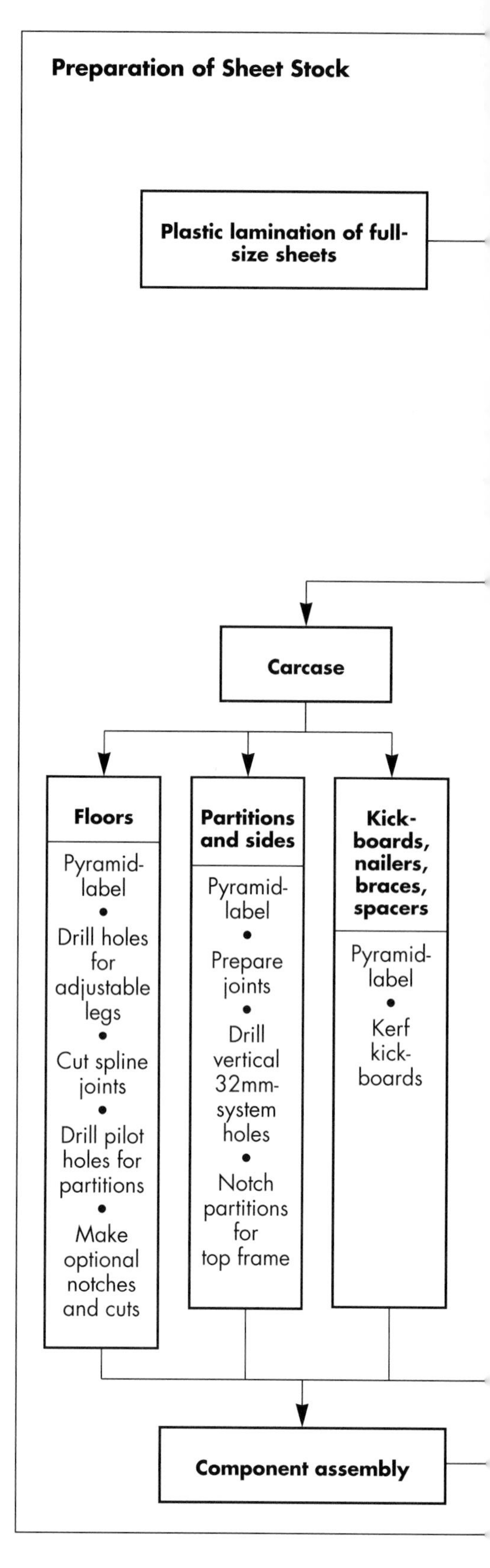

The process of preparing sheet stock is easily the most vigorous task that has to be performed in the construction of a set of cabinets. After sizing and stacking the 20 to 30 4x8 sheets involved in an average-size kitchen, and then performing the various milling operations (such as drilling rows of system holes, drilling shank holes for adjustable legs and slotting for spline-biscuit joints), you'll know you've earned a break. Take one and enjoy it, for a third or more of the project has been completed.

Laminating Sheet Stock

If the project calls for some components to be composed of plastic-laminated sheet stock, you will probably have to laminate the sheets yourself or hire a subcontractor to do it for you, as prelaminated sheet stock is limited in availability and color. I usually use plastic laminate only for kickboards, but occasionally a client will ask for plastic-laminated doors and drawer fronts. In this case, I edge-band the panels with solid-wood strips. If you choose to do the lamination yourself, avail yourself of the method detailed in *Making Kitchen Cabinets* (see Appendix II on p. 143), which enlists the force of gravity to help in aligning the large sheets of laminate with the substrate. Another suggestion: The 3M company makes a water-based contact cement that really works (unlike many other water-based cements, which are relatively weak). It's nonflammable, nontoxic and not cheap, but well worth the money. It's commonly available from

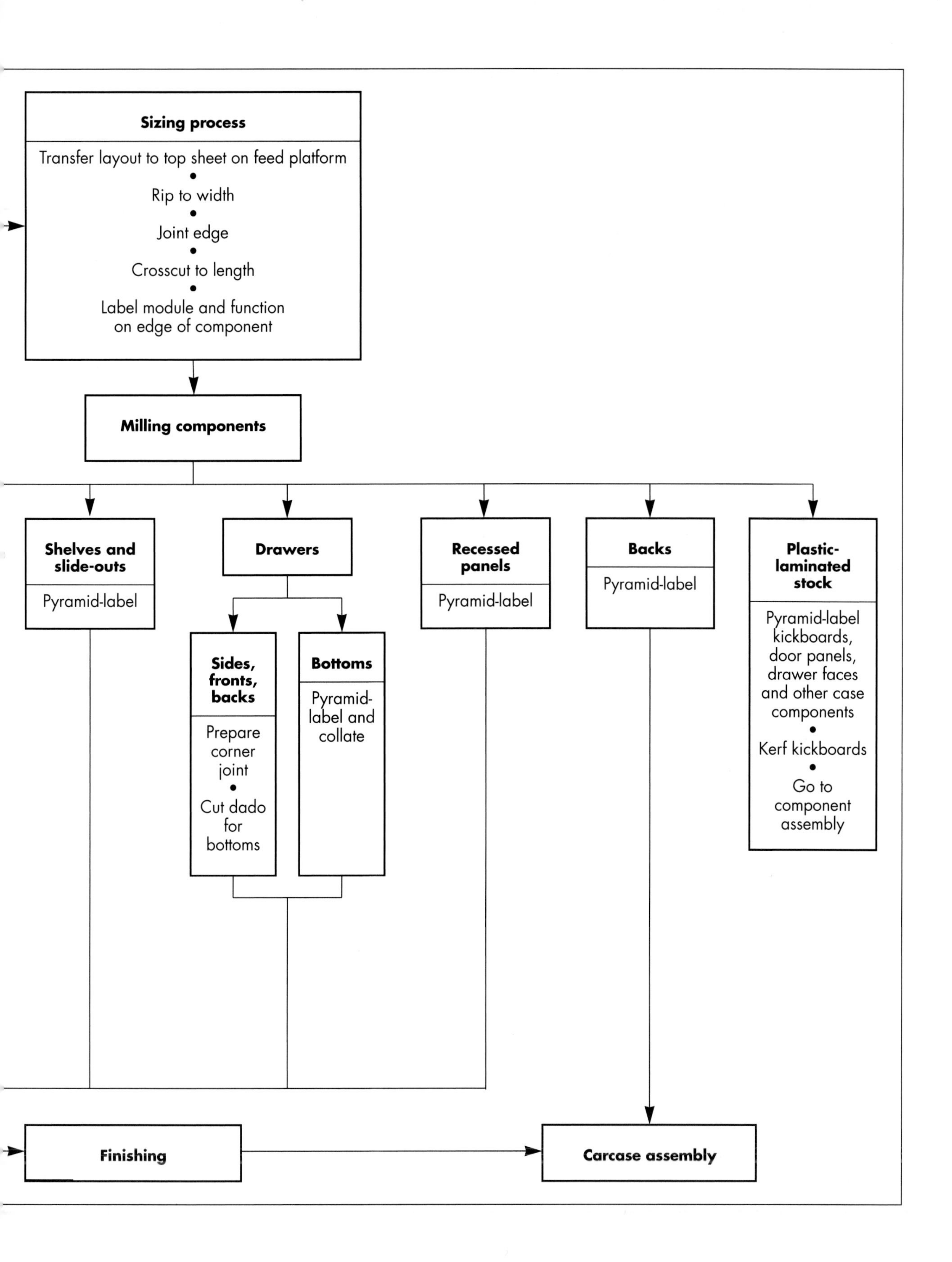
Sizing process
Transfer layout to top sheet on feed platform
Rip to width
Joint edge
Crosscut to length
Label module and function on edge of component
Milling components
Shelves and slide-outs
Pyramid-label
Drawers
Sides, fronts, backs
Prepare corner joint
Cut dado for bottoms
Bottoms
Pyramid-label and collate
Recessed panels
Pyramid-label
Backs
Pyramid-label
Plastic-laminated stock
Pyramid-label kickboards, door panels, drawer faces and other case components
Kerf kickboards
Go to component assembly
Finishing
Carcase assembly

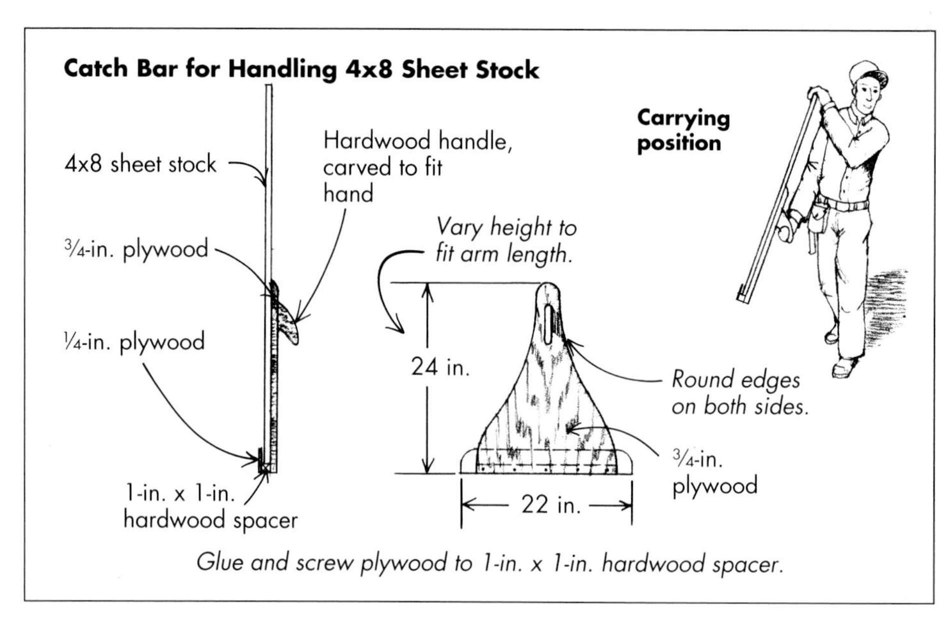

Use a shop-built sliding table to crosscut material that is too wide for the radial-arm saw yet too narrow (under 18 in.) to crosscut safely against the rip fence of the table saw (guard removed for clarity).

material dealers who supply the boatbuilding trade. Otherwise, it can be ordered through any supplier who carries 3M products.

To achieve full bonding strength, the surfaces must be mated with a great deal of force. The force need only be momentary and can be achieved with firm pressure from a rubber roller (somewhat fun) or by rapping a block and mallet over the entire surface (very boring). This is easiest to do with the sheet lying flat, of course, and an ideal location is the 4x8 platform, which happily is where the material must ultimately be located anyway. Thus, after each sheet is laminated (and you need to laminate both sides to maintain the dimensional stability of the stock), bring the sheet to the platform for rapping or rolling. I use a simple, shop-built catch bar (as shown in the drawing above) to make the job of maneuvering sheet stock a cinch. Save your back and build yourself one; it will take less than an hour.

The Sizing Process

With all the sheet stock on the platform, you are ready to begin running the material through the table saw. Well, almost ready. Be sure the plywood blade is mounted on the saw, and run a test piece to ensure that the blade is sharp and not splintering out the underside of the cut, especially on the plastic laminate. Double-check the alignment of the rip fence to the plane of the blade, and adjust the shop helpers so that they hold the stock tightly against the fence without undue drag. (You will have to readjust the shop helpers for different thicknesses of stock.) Referring to your sheet-stock layout drawing (see p. 47), draw layout lines with a piece of chalk on the sheet on top of the pile. You can eliminate this last step once you get into the swing of things and can recall the cutting pattern from a glance at the layout drawing.

Begin by ripping the sheet lengthwise to the largest width specified by the layout, adding ⅛ in. Joint both of the newly cut edges (remember to use the portion of the jointer blade earmarked for sheet stock) and rip each panel again to final dimension, adding ⅛ in. for final planing. If the offcut stock is to be ripped again, rip it to width plus ⅛ in. and final-plane it to size. Jointing the edges of large sheets of material is admittedly awkward, but unless your rip blade yields a cut that will mate perfectly with the applied solid-wood edge band, this step really is necessary. (I have found that a good plywood blade eliminates the need for jointing.)

Now crosscut the 8-ft. long pieces to length. If the run is less than 18 in. wide, I use a sliding table to carry the panel through the sawblade, for safety and accuracy (see the photo on the facing page). This table, which can either be shop-built or purchased, will crosscut the stock into two panels with ends at a true 90° angle to the planed edge. Stock wider than 18 in. can be crosscut safely simply by running a trued factory edge against the rip fence. (Sliding tables built to accommodate these widths are bulky, awkward to handle and often create significant inaccuracy.) First test the factory edge for a true 90° angle with the jig shown in the photo and drawing above. If necessary, proceed to true the edge using this same jig and a fluted bit with a top bearing (see Ap-

Hold the shop-built squaring jig against the factory-cut edge of a piece of sheet stock to check for a perfect right angle.

Jig for Testing and Truing a Sheet-Stock Edge for Square

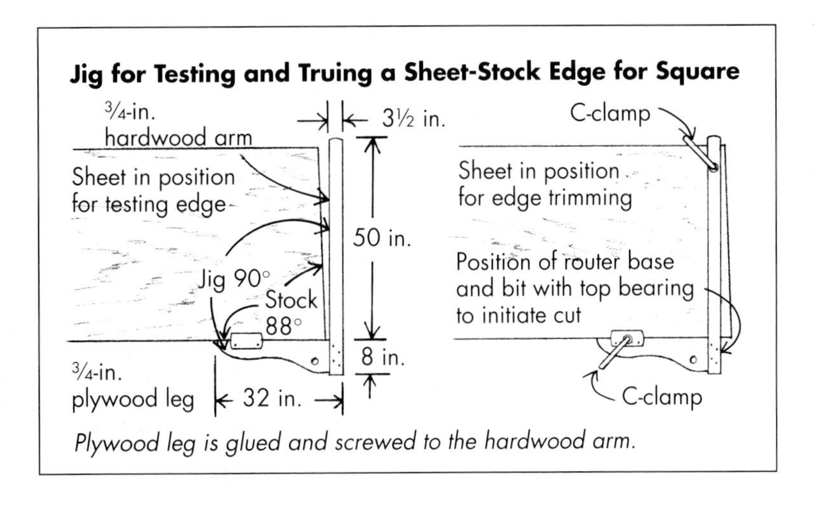

Plywood leg is glued and screwed to the hardwood arm.

If the edge needs trimming, slide the jig back so that 1⁄16 in. to 1⁄8 in. of material is exposed past the jig arm at the narrowest point, and clamp it to the sheet. Trim with a fluted router bit that has a top bearing mounted on the shank above the flutes, using the arm as a guide.

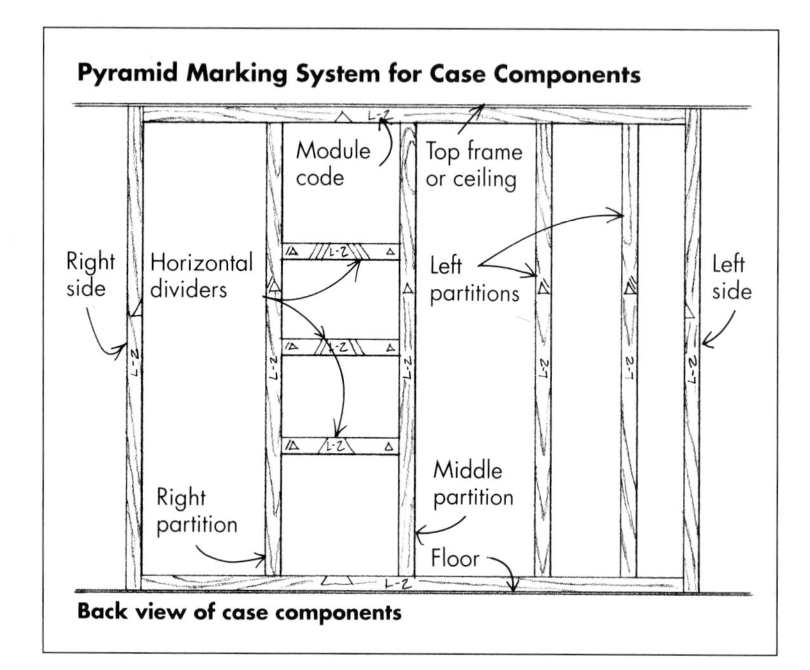

pendix I on pp. 140-142 for a supply source) chucked in a router, as shown in the photo at left. Slide the jig back on the panel to expose approximately 1⁄16 in. to 1⁄8 in. of the factory edge, and run the router along the arm of the jig. Once squared, the panels can be directly indexed to the rip fence and crosscut to exact length.

Continue to rip and crosscut all the sheet stock on the feed platform, stacking the sized panels on Larry, the materials stooge (see p. 32), to be carried to the east wall of the shop. Be sure to write the module designation and function (such as floor, side, etc.) in ink on the edge of each piece. If the stock is less than 1⁄2 in. thick, chalk this information on the surface near an edge. Orient the panels when stacking so that these marks are visible.

Milling Carcase Components

Locate the sized carcase components and organize them by the functions they will serve (such as floors and sides) and stack them on 2x4s placed across two 32-in. lifts. Reorient the 4x8 platform 90° and raise it up to 32 in. (see the top drawing on p. 30) to act as a work surface for the following operations.

Begin with the floor components. As each one is brought up to the platform, orient it the way it will appear in the assembled case and mark its back edge with a portion of a pyramid. Write the module designation beside it. This symbol will orient and identify the panel during the rest of its journey through the production process. If you see a pyramid, or a portion of one, on the edge of a component, you automatically know that it's the back edge; the portion of the pyramid shown reflects the type and orientation of the component. As shown in the drawing at left, the top half of a pyramid indicates a top, while the lower half indicates a bottom. Right or left portions appear on sides and full pyramids are drawn on vertical partitions.

Continuing with the floor components, locate and drill the holes that will receive the bolts for the adjustable legs, using the jigs shown in the drawing below and the photo at right. Use the same jigs to locate the legs along the back of the floor components.

The next step to be performed on the floor components depends on the type of joint you will use to assemble the carcases. If the units are to be of knockdown construction, nothing further need be done, as the pilot holes for the Confirmat knockdown fasteners (see p. 92) are drilled during the case-assembly process. The floor panels can now be stacked on Larry to be brought to the south wall of the shop.

A simple shop-built corner jig locates the holes in the cabinet floors for the adjustable legs. Stops on both sides of the jig allow it to be flipped over and used at the opposite corner. To minimize tear-out, use a sharp brad-pointed bit.

Jigs for Locating Holes for Adjustable Legs

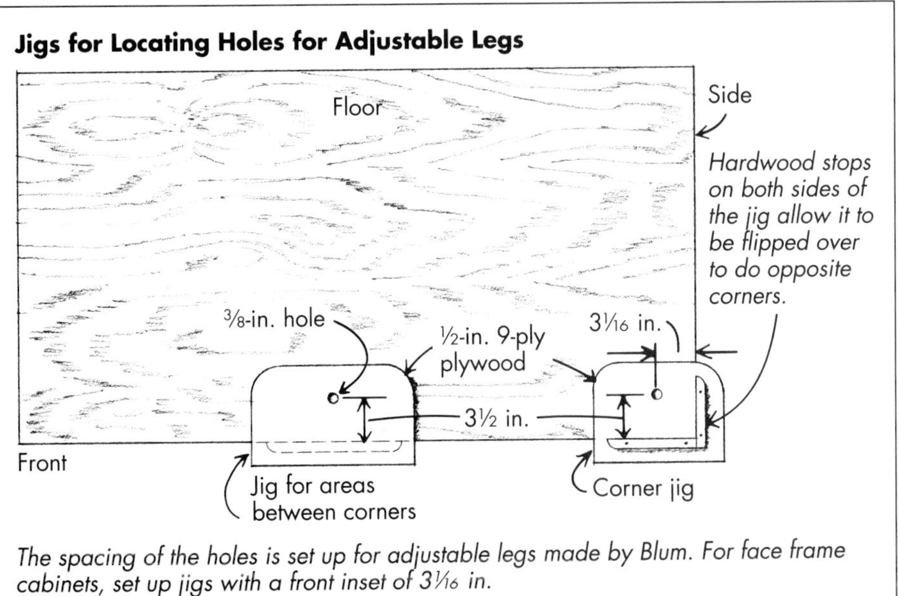

The spacing of the holes is set up for adjustable legs made by Blum. For face frame cabinets, set up jigs with a front inset of 3 1/16 in.

When slotting for a partition intersecting the middle of a floor, the back-to-back clamp functions as a stop for the spline-biscuit machine. The marks made on masking tape applied to the clamp locate the centerline of each biscuit slot.

Here the machine cuts slots for the biscuits in a partition edge. The back-to-back clamp locates the centerlines, while the work surface itself indexes the base of the machine. It's necessary to clamp the panel to the work surface.

If, however, you choose to use spline-biscuit joints, this is the time to lay out and cut the slots. Prepare a layout guide, which indicates the location of the center of each spline and also acts as a stop for the biscuit-joining machine while slotting in the middle of a panel for a partition. Your 36-in. back-to-back bench clamp serves nicely as a guide (see Appendix I on pp. 140-142 for a supply source). All that is necessary to prepare it for this application is to draw lines across the bar and down each side to indicate the centerline of each slot. Then clamp it across the panel along both edges; the center mark on the joining machine is aligned to the marks on the bar clamp, as shown in the photo above. For guiding the machine on partitions, secure the clamp across the panel along an edge and simply run the machine against the clamp at the indicated marks, as shown in the photo at left. Be consistent: Always run the guide along

When running slots along the ends of side panels to receive a floor or top, add a strip of material the same thickness as the side to support the machine.

the left edge of a partition (as seen from the face of the cabinet).

Once the slots for the partitions have been cut, run a pilot hole next to each slot sized for the screws used to draw the components together during assembly. I have found that 1⅝-in. drywall screws are ideal for this application. Except in laminated stock, they require no additional countersinking or pilot drilling to be installed. If any special cuts need to be made in the floors, such as an indentation in front of a sink area, make them now.

When all the floor components have been processed and stacked on Larry, roll them to the south wall of the shop to await edge-banding and finishing.

Now load the side and partition components onto the platform. If the case is to be joined with spline biscuits, continue the slotting process.

To slot the sides to receive floors, ceilings and top frames, clamp the layout guide to the side panel exactly ¾in. (or the thickness of the stock to be joined) away from the edge. As when slotting for partitions, hold the spline-biscuit machine vertically and plunge it down into the side stock. Since the faceplate of the machine bears on only ¾ in. of material, place a strip of stock the same thickness as the side stock along the edge for support, as shown in the photo above. If horizontal partitions will join the case sides, index the clamp guide consistently to the bottom of the partition.

Drilling 5mm holes along the sides of a case is easily accomplished with a commercial jig. This one, by J&R Enterprises, uses two bars, allowing both front and back holes to be set up simultaneously.

The sides are now ready to receive their vertical 32mm-system holes. Commercially made drill guides are available that will precisely align a 5mm brad-pointed bit on 32mm centers (see the photo at left). For the price of a table saw, you can add a line-boring head to your drill press that bores five holes at once; for the price of a really good table saw, you can obtain a freestanding unit that bores up to 13 holes at a time (see Appendix I on pp. 140-142 for manufacturers). A typical system-hole layout is shown in the drawing below.

If the cabinets have face frames, you won't have to drill system holes for European-style hinge plates or drawer slides, as these will attach directly to the sides of the face frame. In this case, you will need to drill the system holes only in the areas that will have adjustable shelves. If, however, the cabinets will not use face frames to carry the door

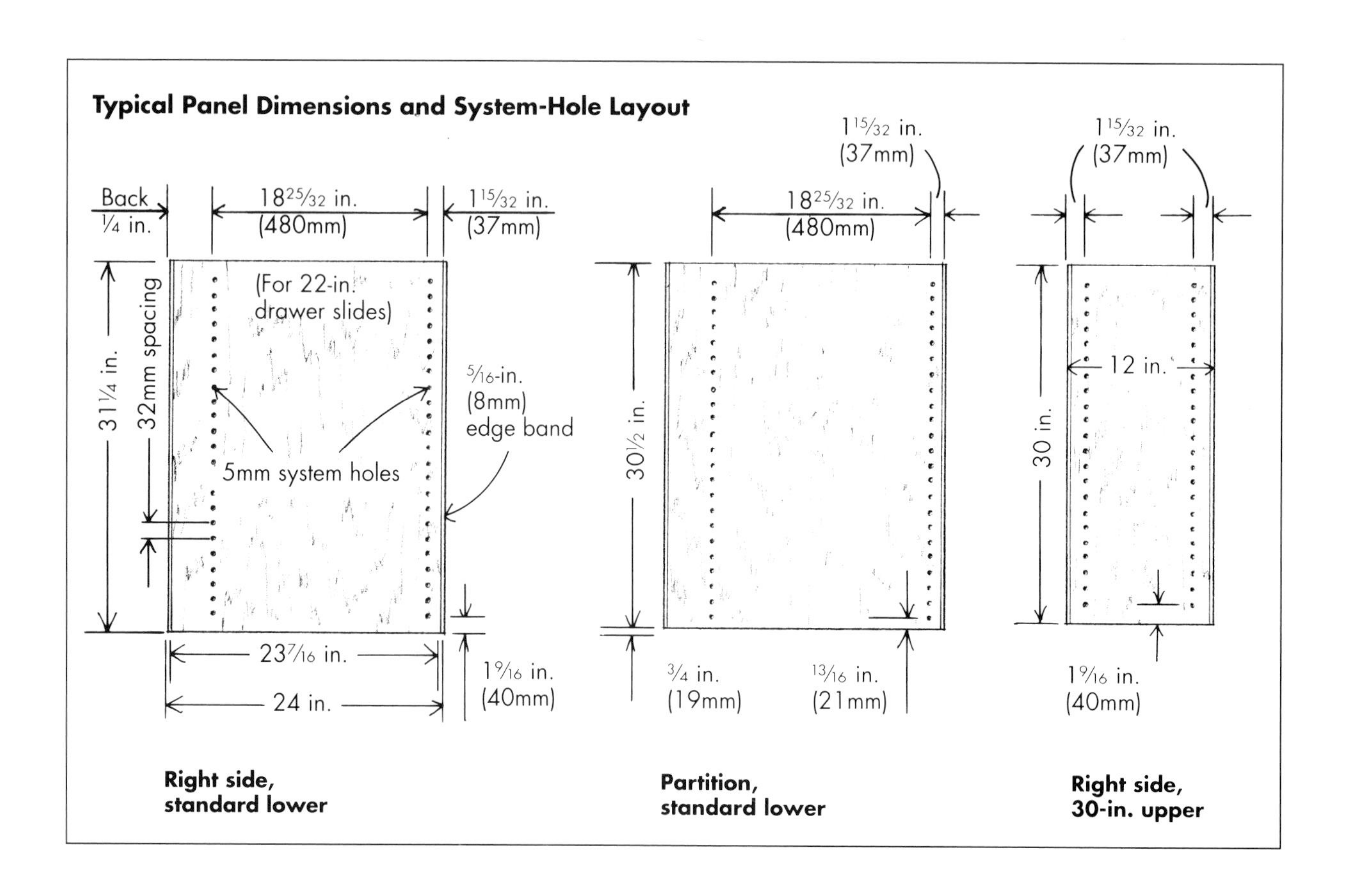

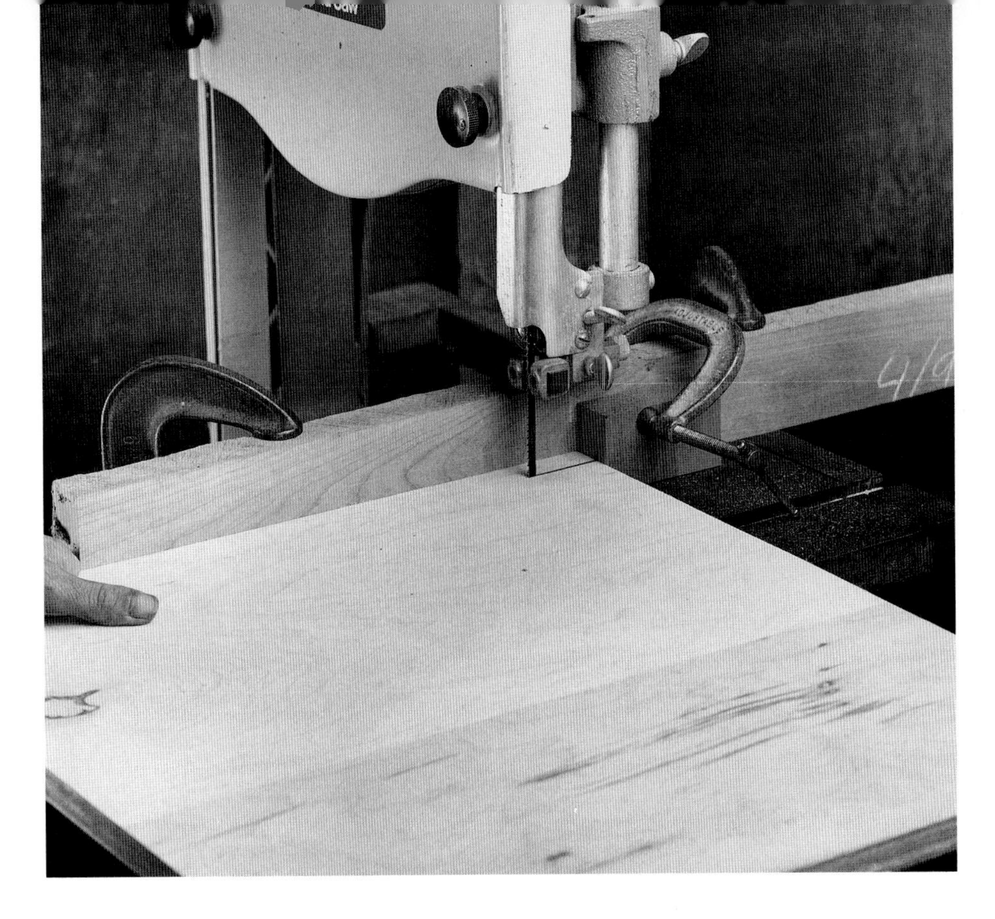

Notches in the partitions are easy to cut on a bandsaw if a fence with a stop is set up to guide the cut. Here, the first cut is being made. Reorient the guides before making the second cut.

and drawer hardware, the sides and partitions must receive their full measure of system holes. For these holes to be properly positioned to receive hinge plates and drawer slides, they must be located a precise distance in from the vertical edge and up from the bottom. In all cases, the holes must be 37mm (1 15/32 in.) on center from the front edge of the cabinet. Remember to take into account the 5/16-in. (8mm) edge band that still has to be applied. The spacing from the rear edge of the panel is a function of the drawer-slide length (hole spacing is standardized on slides throughout the industry) and the depth of the panel. For example, 22-in. drawer slides provide for a rear screw setback of 480mm (18 25/32 in.) from the front screw. Simply lay out this spacing and then measure in from the back edge of the panel. The adjustable stops on the drill guide are set to the inset required; the base of the guide is always indexed to the bottom of the sides.

Jig partitions to receive the system holes in exactly the same way as the sides, except temporarily add a spacer of the same thickness as the floor stock to the base of the partition to catch the drill guide's bottom stop and maintain alignment. If notches are to be cut in the partitions, cut them now with a jigsaw or bandsaw. The bandsaw is faster and produces a more accurate cut (see the photo above). Load the completed side and partition components onto Larry, the materials stooge, and stack them with the floor components. Label the remaining carcase components (nailers, spacers and braces) and bring them to the south wall as well. Kerf the kickboards with a saw cut to receive the clips needed to mount them on the adjustable legs, and then stack them with the nailers, spacers and braces.

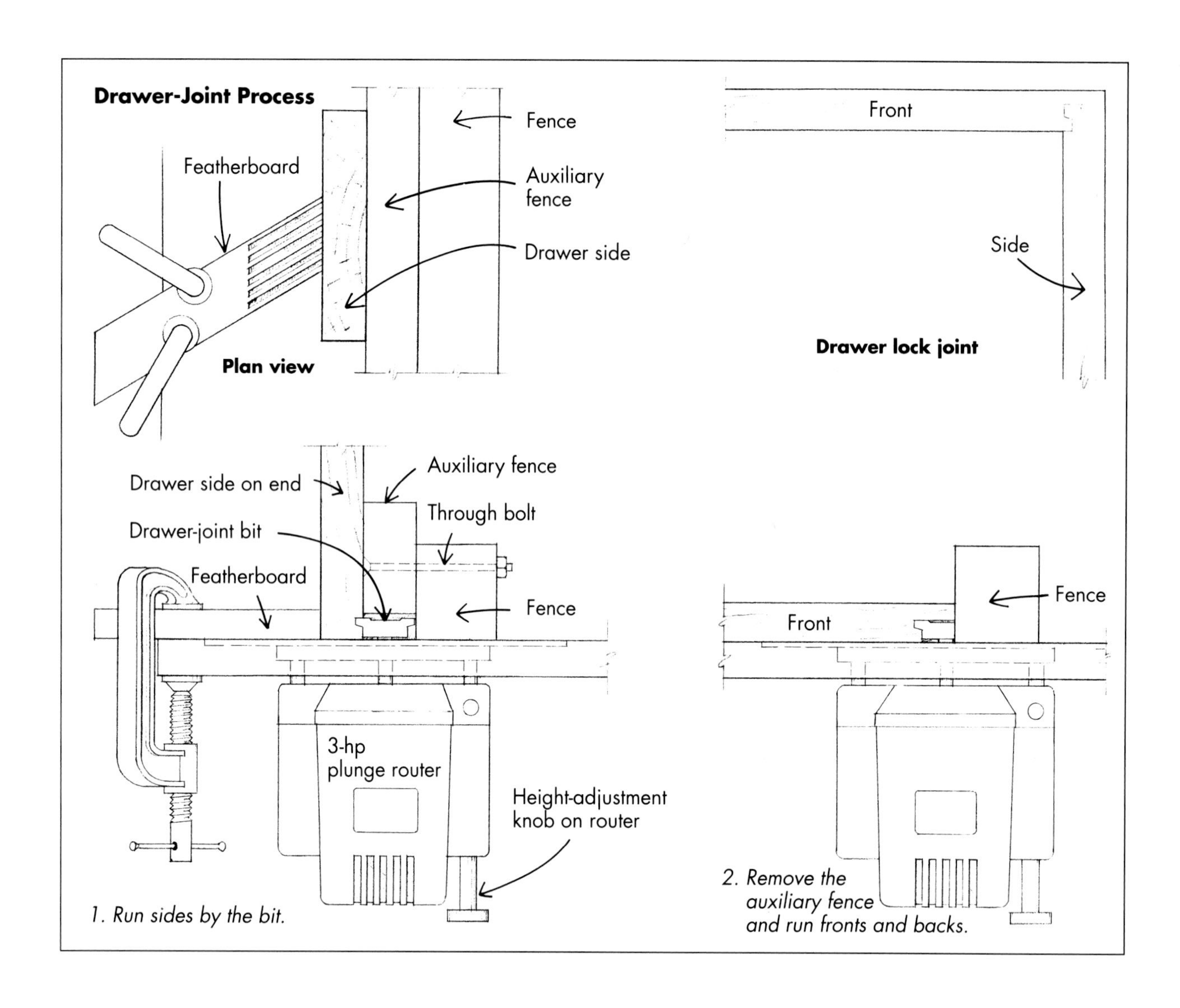

Milling Other Sheet-Stock Components

The stock appearing next in the stack against the east wall should be the ¾-in. thick material for shelves and slide-outs. Label these components and bring them to the south wall to await further processing (edge-banding for the shelves; assembly with solid-side components for the slide-outs.)

Next in the stack are the ½-in. thick drawer parts. Sort these components into two piles—one for drawer sides, the other for fronts and backs—and place them on the stooge. Wheel them to the shaping station and set up the router with a drawer-joint bit. I use the bit made by Freud (see Appendix I on pp. 140-142); it's inexpensive, yet makes a strong joint. Shape a number of scrap pieces until you are satisfied with the joint. Rather than moving the fence to shape the joint on the drawer fronts and backs, use an auxiliary fence when routing the drawer sides, as shown in the drawing above. Secure a featherboard to the table surface to hold the components tightly against the fence, back the pieces with scrap (to avoid splitting) and push through.

Before leaving this work station, remove the drawer-joint bit and chuck in its place a slotting bit to cut the dado for the drawer bottoms. Make a sample cut to test that the inset from the edge and the depth of the dado are satisfactory (I use 5⁄16 in. for both), then cut all the joints. Because the side of the component to receive the dado is always the side that has received the drawer joint, prepare the stock for an efficient feed process by orienting all the faces in the same direction. Once the dadoes have been cut, wheel the components to the south wall of the shop and stack them away from the carcase components and shelves.

The last panels left in the east wall stack (unless plastic-laminated components are involved in the project) are the ¼-in. thick drawer bottoms, recessed panels (if applicable) and carcase backs. Sand the drawer bottoms to 180 grit before labeling them and moving them to the stack of drawer components on the south wall. Sand the recessed panels with 220 grit and then stack them in the area just to the left of the drill press. The case backs, assuming you are using a vinyl-coated material, need only be labeled and then stacked with the rest of the carcase components.

Milling Laminated Panels

Depending on the job, face components, carcases and kickboards will sometimes be made from plastic-laminated stock. The sized components that will become doors need to be edge-banded and drilled to receive cup hinges. Because this occurs in later production blocks, separate these components from the other sized panels and hold them in a separate stack against the east wall. Also find the drawer fronts, which will require edge-banding as well, and bring them to the south wall of the shop to await that process.

Any strips of laminated stock earmarked for kickboards are run through the table saw to receive a ⅛-in. kerf along the length of the back. This kerf holds the press-in clips that attach the kickboard to the adjustable legs. The completed kickboards are stored along the south wall of the shop; they'll be cut to length and installed during the carcase-assembly production block (or during installation on projects that require the kickboards to span more than one module).

Any laminated panels destined to become carcase components are pulled from the stack and brought up to the 2x4s set up by the work platform. Refer to the elevation cards to determine which module and function each panel serves. Pyramid-label the components and write the code of the module on their edges. Perform the appropriate operations, and then stack the panels along the south wall.

Chapter 10: Preparing Solid Stock

There was a time in my shop when solid-stock preparation was more than a production block; it was, in fact, almost the entire process. In addition to doors, drawer faces and face frames, I also built case sides, shelving, drawer boxes and even countertops of solid wood. Only cabinet floors and backs were made of sheet stock.

The tedious processes involved in working with solid stock filled my shop with fragrant aromas, but they did little to fill my pockets with cash. These days, as you've just seen in the previous chapter, I make up much of the casework from sheet stock. Solid stock, however, remains the material of choice for doors and drawer faces, applied end panels, face frames and components such as moldings, exposed nailers, edge bands and top frames of lower cabinets. Accessories including door-mounted bins, appliance garages and exposed shelving are also best made up of solid stock, and further contribute to the sense of warmth and quality of the cabinetry.

Surface Planing

When you order solid stock from your suppliers, ask them to plane it to ⅛ in. over the finished dimension you need, and to leave the edges unjointed (this stock is referred to in the trade as S2S, that is, surfaced two sides). Doing the final preparation yourself is the only way to ensure that you will have smooth, uniform material to work with.

Prepare for planing by moving the 4x8 platform to the southern half of the shop and setting it on the 32-in. lifts. Pull the surface planer off the wall or down from the ceiling and set it in the middle of the platform, running lengthwise. Don't forget to hook up the vacuum hose to the dust chute. Round up Larry, the materials stooge (see p. 32), and load up the stock stored on the first rack above the radial-arm saw (this is typically the 1-in. stock planed by your supplier to ⅞ in.). Wheel the stock to the platform and position it down to the right of the planer as you face the infeed. If you need more room for the stock than is available on the platform, insert 8-ft. long 2x4s crossways between the platform and the planer support grid.

For the smoothest possible surface, set the planer to cut off only 1⁄32 in. of wood at each pass. Verify that the knives are sharp—if they're not, take 20 minutes and hone the knives on your waterstones. (Ryobi provides a honing jig with its planer and another jig for precisely resetting the knives in the cutterhead.)

When everything looks good, feed the stock through and stack it to the left of the planer. Lower the cutterhead to remove another 1⁄32 in. and feed the material through once again, flipping the boards to expose the opposite surface to the blades. Repeat the process until you have the final dimension. I surface 1-in. stock down to 13⁄16 in. This allows me to shape more definition into the molded edges of doors and drawer faces than would be possible in ¾-in. stock. Another advantage is that edge bands can be ripped from the stock in one operation.

Layout

I always anticipate the layout stage with some trepidation. Faced with a pile of boards of varying widths and lengths, the task is to discover the most productive way to lay out all the components that must be cut from the wood.

I try to eliminate as much material as I can in as short a period of time as possible (before my brain cells overheat). Begin by sorting through the pile, culling out the widest pieces of stock. Spread these out across a pair of 32-in. lifts, then refer to the solid-stock cut list. Since the list is broken down by widths, it is a simple matter to assign the widest components to the widest stock. Chalk the layout lines directly on the stock, allowing at least 1 in. waste at the board ends. Take care to avoid defects in the wood. Many of the wider components are drawer faces; laying out adjoining drawers on the same board will maintain a pleasing flow of grain patterns from one cabinet to the next. A bank of drawers can be laid out and grain-matched by edge-laminating enough stock to form a panel 1 in. longer and 1 in. wider than the combined height and width of the faces. This panel will be ripped to the required individual widths after surfacing (see p. 75).

The boards to make up any other edge-laminated panels called for in the project will also be cut from the wide stock. You will find that a significant amount of board footage rapidly becomes committed, and the task begins to take on a more manageable proportion.

Select door rail and stile stock from the straightest wood left in the pile. If there are any paired doors, try to lay out adjoining stiles on the same length of stock, then rip the stock into two pieces. This allows the grain to match perfectly from door to door, and any movement in the wood will occur in a similar way in each stile.

Whenever possible, try to group components of similar length on the stock, so that you will be able to rough-crosscut them to length in the same operation. This is especially helpful with stock that has a significant amount of curve to the edge. The board yields more width if it needs straightening only over portions of its run, as shown in the drawing below. Moldings, edge bands and nailers should be laid out to run the full length of the stock. Note the straight reference lines, which indicate where to begin the first rip.

Repeat these layout steps for each thickness of stock used in the job.

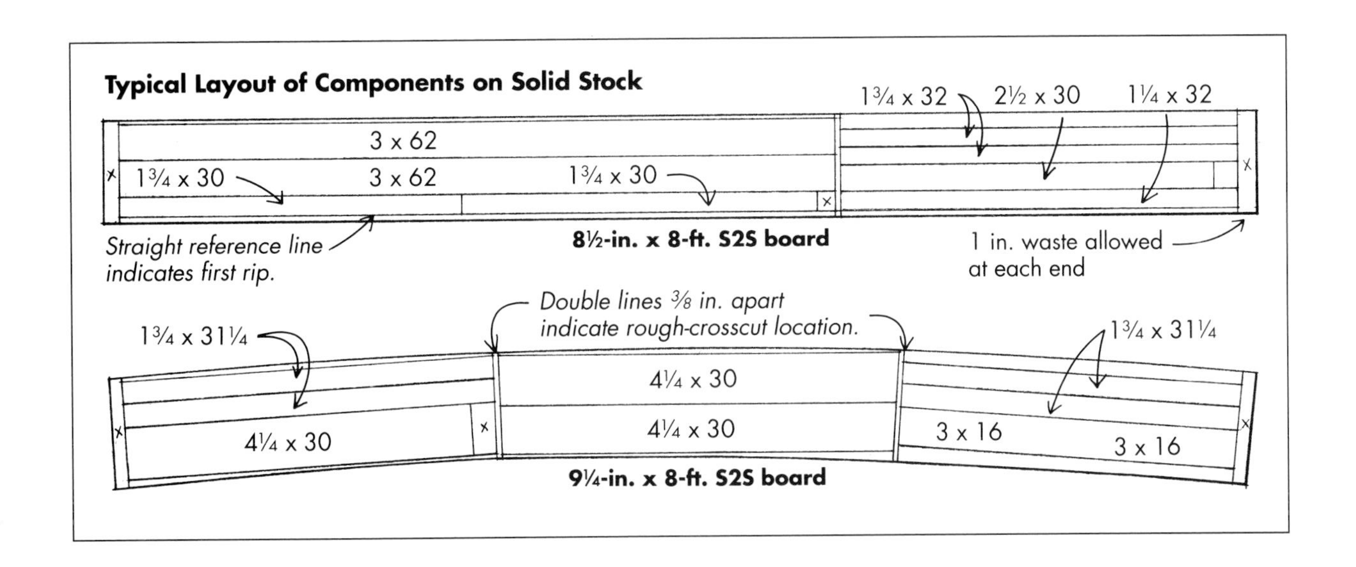

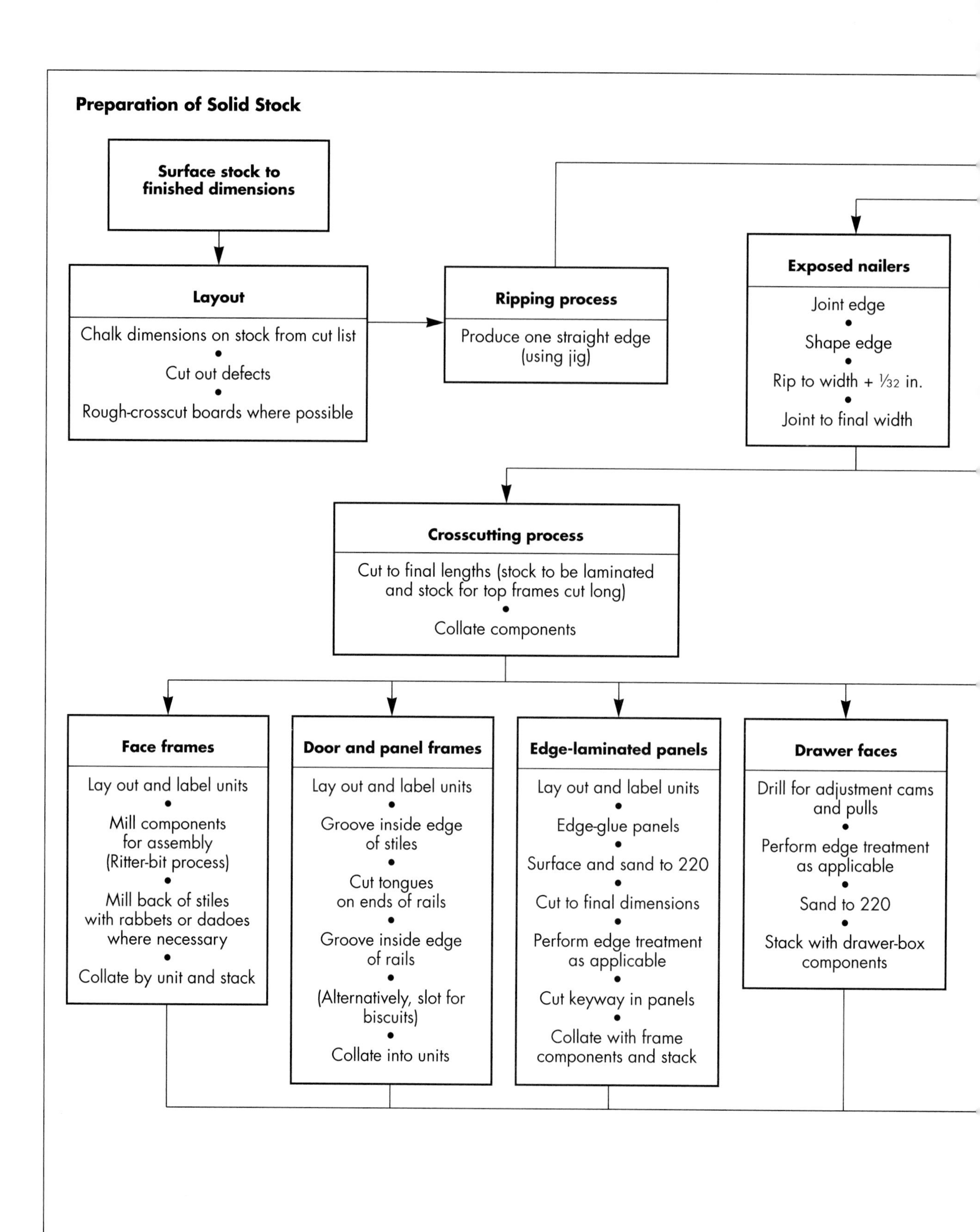

Preparation of Solid Stock
Surface stock to finished dimensions
Layout
Chalk dimensions on stock from cut list
•
Cut out defects
•
Rough-crosscut boards where possible
Ripping process
Produce one straight edge (using jig)
Exposed nailers
Joint edge
•
Shape edge
•
Rip to width + 1/32 in.
•
Joint to final width
Crosscutting process
Cut to final lengths (stock to be laminated and stock for top frames cut long)
•
Collate components
Face frames
Lay out and label units
•
Mill components for assembly (Ritter-bit process)
•
Mill back of stiles with rabbets or dadoes where necessary
•
Collate by unit and stack
Door and panel frames
Lay out and label units
•
Groove inside edge of stiles
•
Cut tongues on ends of rails
•
Groove inside edge of rails
•
(Alternatively, slot for biscuits)
•
Collate into units
Edge-laminated panels
Lay out and label units
•
Edge-glue panels
•
Surface and sand to 220
•
Cut to final dimensions
•
Perform edge treatment as applicable
•
Cut keyway in panels
•
Collate with frame components and stack
Drawer faces
Drill for adjustment cams and pulls
•
Perform edge treatment as applicable
•
Sand to 220
•
Stack with drawer-box components

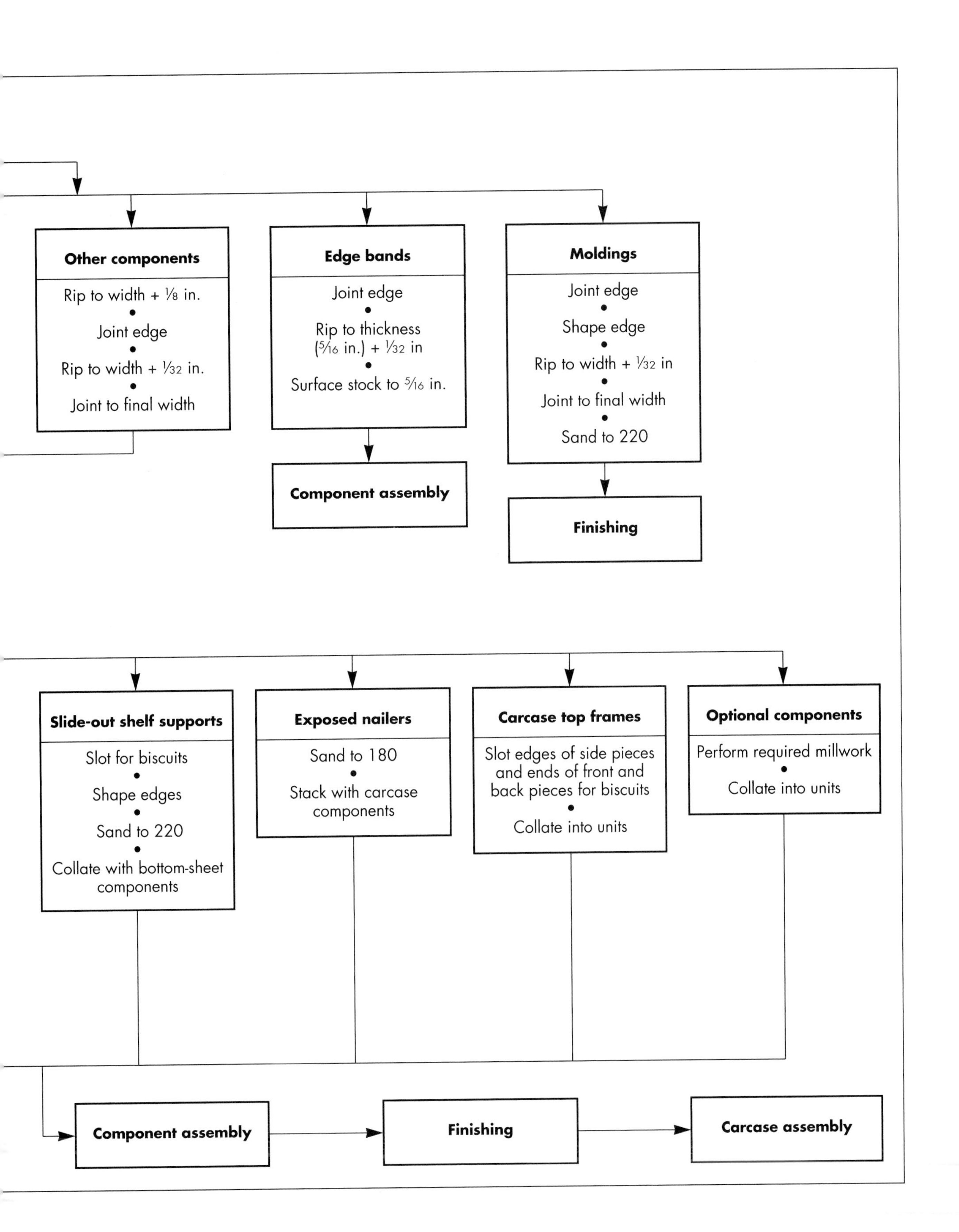

Other components
Rip to width + 1/8 in.
Joint edge
Rip to width + 1/32 in.
Joint to final width
Edge bands
Joint edge
Rip to thickness (5/16 in.) + 1/32 in
Surface stock to 5/16 in.
Component assembly
Moldings
Joint edge
Shape edge
Rip to width + 1/32 in
Joint to final width
Sand to 220
Finishing
Slide-out shelf supports
Slot for biscuits
Shape edges
Sand to 220
Collate with bottom-sheet components
Exposed nailers
Sand to 180
Stack with carcase components
Carcase top frames
Slot edges of side pieces and ends of front and back pieces for biscuits
Collate into units
Optional components
Perform required millwork
Collate into units
Component assembly
Finishing
Carcase assembly

Ripping the First Straight Edge

Prepare for the ripping process by rough-crosscutting the boards to length where possible. Rather than hauling them over to the radial-arm saw, I just crosscut the boards with a jigsaw as they sit on the lifts.

To avoid free-handing the material through the table saw, I rip the first straight edge on each board using the method shown in the drawing at left. Make sure the edges of the hardwood straightedge are perfectly straight and parallel. Rip the board by running the straightedge against the rip fence, with the blade set to cut just to the left of the reference line.

Creating a Straight Line on Rough Solid Stock

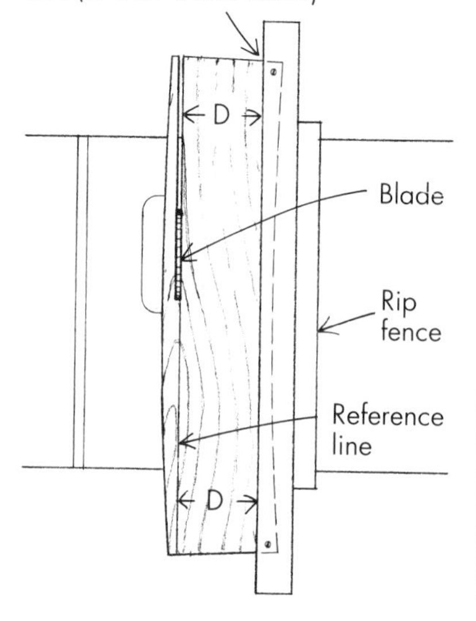

The straightedge is located on the stock so that D = D. The right edge of the stock must not protrude beyond the right side of the straightedge.

Ripping Edge Bands And Moldings

Now sort through the lumber pile and cull out all the boards designated for edge bands and moldings. Ripping this stock is a little different from processing the other components.

Begin with the edge bands. Take the first board and run its ripped edge over the jointer, removing just enough material to get rid of the saw marks. (It isn't critical that the board be jointed perfectly straight.) If the stock is thicker than the ¾-in. thick sheet stock it will have to band, you will only need to rip it once at 5⁄16 in. plus 1⁄32 in. to produce an edge band. (If it is not thick enough to provide an overhang for trimming, rip the stock twice, once at 13⁄16 in. and then again in half to produce two 5⁄16-in. plus edge bands. You may need to use a thin-kerf rip blade or a bandsaw to get two 5⁄16-in. plus pieces out of ¾-in. stock.)

Use the shop helpers to hold the stock down on the table tightly against the rip fence. Use a push stick as shown in the photo at left to move the edge-band offcuts through the blade; you will have to keep the push stick parallel to the table to get under the arms of the shop helpers. As each edge band is cut off the board, joint the rough edge of

If you're using shop helper hold-downs on a rip fence, push the material through the sawblade with a long hardwood stick that is narrower than the width of the cut. Keep the stick parallel with the table and hold it against the fence.

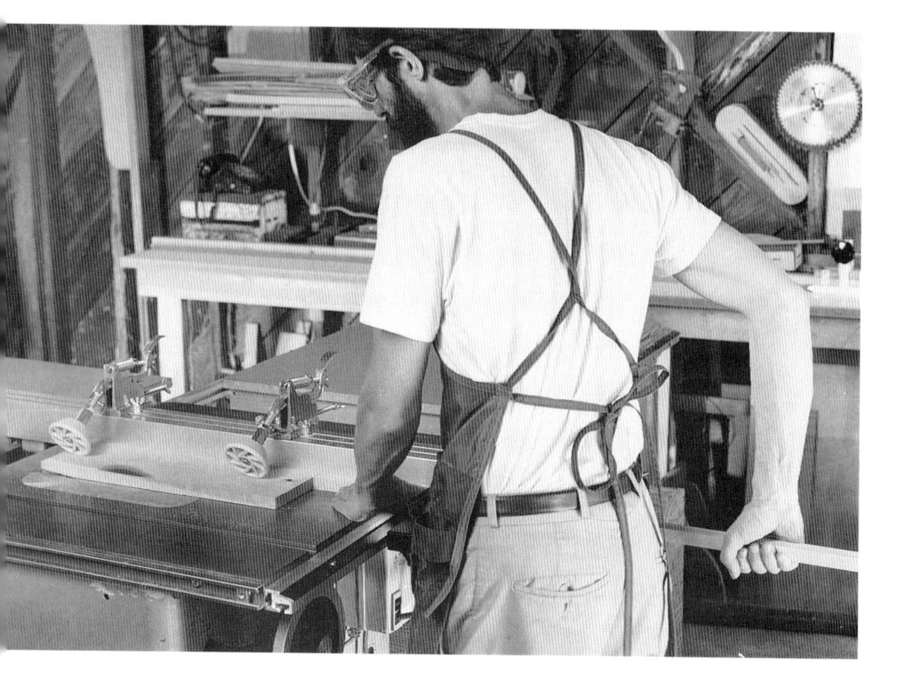

the parent material and rip out another edge band. Continue until you've milled all the edge bands.

Let the jointer, table saw and dust collector run continuously throughout this operation. This saves wear and tear on the motors, which have to strain against inertia every time they are asked to start up. It also saves a little time in fumbling for switches, and it fills your shop with the sweet sounds of progress.

The next step is to run the bands through the surface planer to true the stock to exactly 5⁄16-in. thickness. Stack the completed edge bands with the carcase components along the south wall and hang the planer back on the wall (or haul it up to the ceiling).

Preparing moldings and exposed nailers (which can be seen in the top rear corner of upper cabinets) is a similar process to that used for the edge bands, except that it is necessary to shape the edges of the stock between jointing and ripping. Thus, yet another machine is added to the symphony of progress. Here are the steps. Smooth the rough edges of the parent stock on the jointer. Run the smoothed edge by the shaping cutter. Rip the molding/nailer to width plus 1⁄32 in. and set it aside. Repeat until all moldings and exposed nailers are complete.

Run the pieces over the jointer to remove the extra 1⁄32 in., using push pads with nonslip surfaces, not your hands, to move the stock past the cutterhead. Sand the moldings to 220 grit and set them up out of harm's way on the lumber racks until it's time to finish them. Stack the exposed nailers on Larry in preparation for crosscutting.

Ripping the Other Components

The balance of the stock for face frames, carcase frames, door and panel frames, drawer faces and edge-laminated panels is ripped to width in a similar fashion. Begin with the wider stock and work your way down to the narrowest components. Use the hardwood straightedge shown in the drawing on the facing page to rip the initial straight edge on each board. Then rip to the required width plus 1⁄8 in. (Don't bother jointing the edge between subsequent rips.)

Once all the stock has been processed, sort the material into piles based on width. Beginning again with the widest stock, joint any bow out of the board (which is why you provided that 1⁄8 in. of extra width). If the board is so bowed that jointing will decrease the width below the desired dimension, rough-crosscut the stock at the layout marks with your jigsaw and then joint the remaining bow out of the edge. Proceed to rip the stock to final width plus 1⁄32 in. Finish the operation by jointing off this 1⁄32 in. from the rough edge. Continue the process until you've dimensioned and stacked all the boards on Larry, and then wheel the load over to the east wall of the shop between the radial-arm saw and the drill press. Leave the material on the stooge for now.

The Crosscutting Process

Bring the 4x8 platform and lifts over to where you have parked Larry. Sort the material by length, removing the boards containing the longest components to be cut and stacking them to the left of the platform. Continue to sort until all the material has been transferred from Larry to the platform. The last pieces stacked to the right side of the platform should contain the shortest components. Roll the empty stooge near the radial-arm saw to receive the components as they are cut.

Now set the sliding stop on the radial-arm saw fence to the greatest length required and begin crosscutting. Work your way through each stack toward the shortest lengths. Cut all components to the exact dimension specified by the layout. The exception is stock to be edge-laminated into panels or banks of drawer faces and stock for top frames: These components are cut 1⁄2 in. long.

A shop-built jig holds a face-frame component at 22½° from vertical while a low-angle countersink bit prepares the piece to receive screws for assembly.

When the last part has been cut, roll Larry next to the platform again and re-sort the parts, separating them by function: face frames, door frames, fixed panel frames, carcase top frames, exposed nailers, slide-out shelf supports, drawer faces and panel stock.

Preparing Face Frames

Remove everything but the face-frame parts from the platform, and stack in their respective piles against the shop's east wall. Now, using the elevation cards, sort the face-frame components into modules, orienting and labeling the pieces using the pyramid system (see p. 54). Keep in mind that face frames often extend over more than one module. Position the piles to the north end of the 4x8 platform; if more room is needed, use Larry's top shelf.

If a project calls for face frames, I find the simplest, fastest way to make the frame joints is with a Ritter bit, a long countersink bit designed to drill at a low angle to the surface of the wood. (See Appendix I on pp. 140-142 for suppliers.) The joint is then glued and fastened with special screws: The juncture of the shank with the head is absolutely square, and thus will not split out the end grain. (The bit suppliers also carry the screws.) The resulting joints are very strong, require little surfacing to become flush and, best of all, require no clamping while the glue dries.

In addition to the Ritter bit, you will also need a jig to hold the pieces in position for drilling. You can either buy the jig or make your own; the photo at left shows the homemade jig in place on the drill press. The support for the stock to be drilled is built 22½° away from vertical and has lines drawn on its surface 90° to the base of the jig. These lines indicate the proper alignment of the stock for drilling.

To prepare for countersinking, chuck the Ritter bit into the drill press and set the speed at about 1000 rpm. Set the jig on the table surface in approximate position and lightly clamp it down. Run a series of test holes in scrap the same thickness as the face-frame stock until the exit hole is centered in the end of the stock and the countersink hole is deep enough to bury the head of the screw. When you are satisfied, set the depth stop for the quill travel and tighten down the jig clamps.

Begin the drilling process by orienting the face-frame components into the configuration shown on the elevation card for the module. To indicate where to drill, mark the back surface of each end that butts into another. Load the marked components onto Larry and roll them to the drill press. Drill the holes where indicated. Pieces 1¼ in. to 2¼ in. wide receive two holes, while narrower stock receives only one. Wider stock can receive a hole every ¾ in. to ⅞ in.

While drilling the holes, be sure to hold the stock perpendicular to the drill-press table; the lines on the jig serve as a reference. Once drilling for the module is complete, integrate the drilled pieces with the rest of the module's pieces, which were left back on the 4x8 platform. (Stiles that abut walls can receive a rabbet to make fitting easier; this should be done at this time.) Wrap a piece of tape around the pile to hold it together and place it to the right of the drill press. Repeat this process for each face-frame unit.

Preparing Door and Panel Frames

Bring all the boards for the door and panel frames up to the 4x8 work platform and sort them into units as specified on the elevation cards. Lay each unit out in its final configuration, using the pyramid system to orient and identify each component.

Subsequent steps will depend on the style of door and panel called for in the project. I offer two types of framed-panel door: The first uses a variety of sticking (grooving) and coping router bits to create molded tongue-and-groove joints (the dado for the panel is cut at the same time). The second type uses the spline-biscuit machine to create the frame joints and applied moldings to contain the panel. This latter method allows me to create unique doors, as I can develop my own bolection moldings to trim the inner edge of the frames.

For the production of either type of door, the first step is to sort the frame components into two stacks: rails (the horizontal pieces) and stiles (the verticals). If the tongue-and-groove joint is to be used, bring the stiles to the shaping area and stack them on the table saw's outfeed table. Install the sticking bit in the router and run test samples until you are satisfied with the placement and depth of the groove. (This is another good time to use shop helpers.)

When the setup is ready, run all the inside edges of the stiles by the cutter. (The rails will also receive this cut but, to prevent tear-out, not until after the tongues have been cut.) Bring the completed stack of stiles back to the 4x8 platform and gather up the rail stock. Stack this stock on the outfeed table, install and tune up the coping bit and run the ends of each rail by it. Use a square piece of solid stock as a push stick to back the rails through the cut, thereby preventing tear-out and keeping the stock square to the fence (as shown in the photo below).

When all the rails have been coped, reinsert the sticking bit. Tune it up again on scrap, and then process the inside edge of the rails. Bring the completed stock to the 4x8 platform and sort the rails and stiles into their respective units. Tape them together and stack them on the east wall to the left of the drill press.

A push stick and the shop helpers are essential accessories while performing the coping cut on the router.

The stationary slotting tool made by Delta is used to prepare frame components for biscuit joining. Here, the end of a rail (top) and a stile (bottom) are slotted while the pieces are held against a wood stop clamped to the machine's table. Note that the quick-release hold-down secures the work to the table surface, while pressure on a foot pedal feeds the machine into the piece.

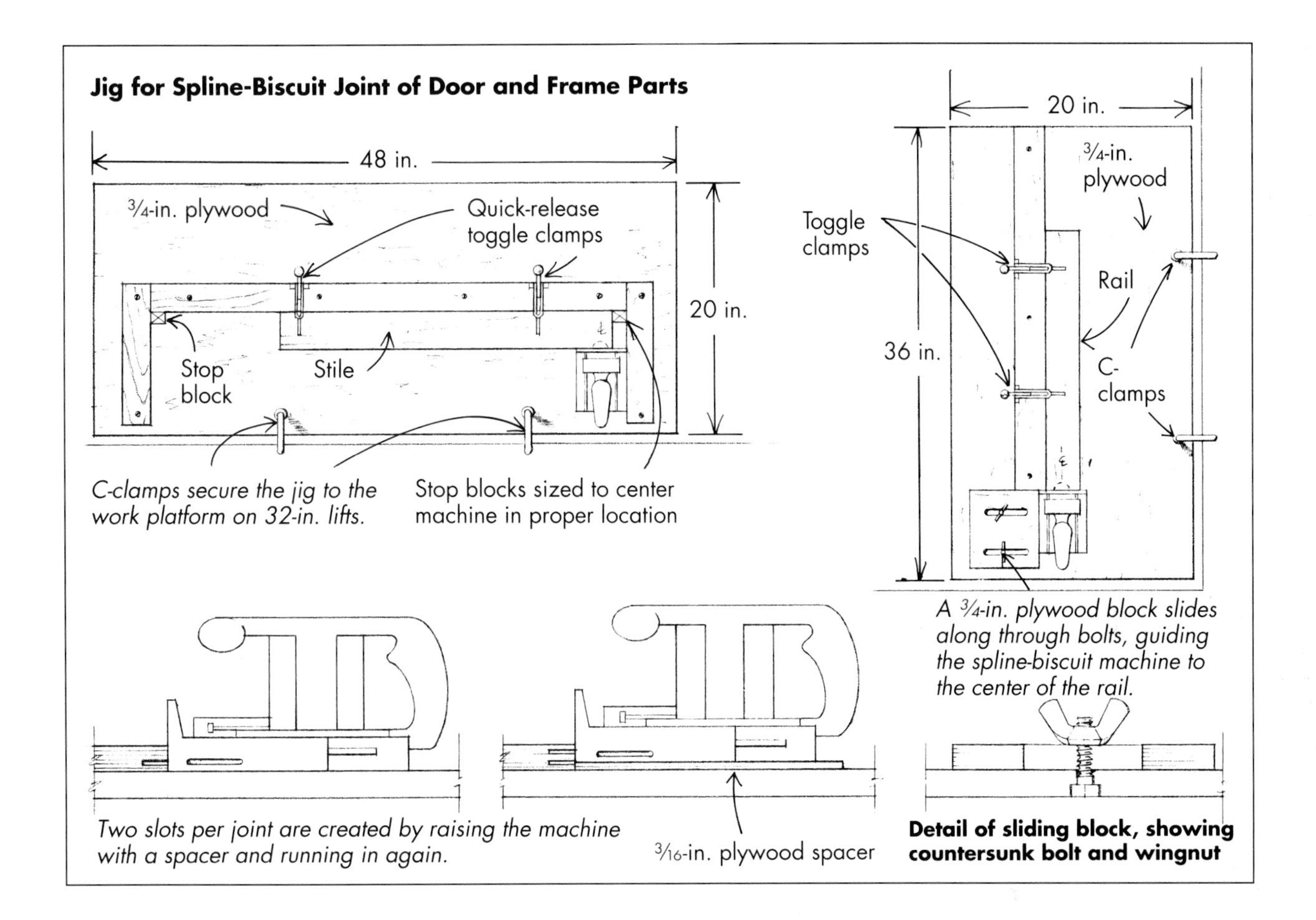

The second type of door, using joints made with spline biscuits, is processed on the 4x8 platform. I use the stationary biscuit joiner made by Delta, shown in the photos on the facing page, to perform this operation. You can, however, use a hand-held biscuit cutter if you make a simple jig to hold the rails and stiles in position. Doubling the splines at each joint increases the strength of the connection. To do this with the hand-held cutter, first run the machine into the wood with the base resting on the jig's work surface. For the second cut, rest the machine on a 3/16-in. spacer, as shown in the drawing above. The stationary machine simply requires that the table be reset to provide a second slot at a different level. Once all the stock has been slotted, sort and bundle it into units and stack along the east wall.

Preparing Panels

Next, bring the stock to be edge-laminated into panels up to the 4x8 platform for sorting and marking. Make sure to include panel stock from which a bank of drawer faces will be created. Align the boards with an eye to creating a pleasing match of grain across the panel's width. I have found that if the stock has been properly dried, that is, if it has about 6% to 8% moisture content, you can safely ignore the orientation of the stock in regard to the direction of the growth rings on the ends of the boards. When you are satisfied with the appearance of the panel, mark a large pyramid on the surface that spans all the boards and write the module code in the center.

I use the setup shown in the photo on p. 72 to hold pipe clamps for the

A 2x6 board, drilled with 1¼-in. holes spaced 8 in. apart along its centerline, is ripped in half to create two 2x3 pipe-clamp supports. The resulting half-circles hold the clamps parallel and level with each other.

edge-lamination process. Rip out two clamp supports from a 2x6 board in which a line of 1¼-in. holes has been drilled every 8 in. along its centerline. Put these supports across a pair of 32-in. lifts. The half-circles securely cradle the pipe clamps, keeping them parallel and level to one another (the pipes should all be of the same diameter).

Prepare to glue up the stock by laying down all the pipe clamps that will fit onto the supports. Distribute the panels along the clamps. If certain lengths of panel fall short of the 8-in. spacing of the pipes, don't center the panel, but keep a clamp under one end and add a clamp to the top of the panel at the other end. To prevent very wide panels from buckling, lay clamps over the top.

Go to the first panel and roll glue onto the surfaces to be joined with a self-spreading glue applicator (for a supplier, see Appendix I on pp. 140-142). Don't overdo it—an overly thick coat of glue will just squeeze out under pressure. You know you have used the perfect amount of glue when it beads uniformly along the joint under pressure, yet does not drip significantly. Apply clamping pressure starting at one end and work toward the other, aligning the board's surfaces as you go along. Be careful not to overtighten the clamps, or all the glue will run out of the joint. If you suspect that certain panels will give you trouble because of opposing deflection in the adjoining boards, use spline biscuits to align the pieces.

Chalk the time on each panel as it is glued up, continuing to laminate until you run out of pipe clamps. If your shop is warm and you have just the right number of clamps, you may experience one of those sweet little moments in cabinetmaking when everything goes just right: The first panel will be ready to unclamp just as you have used up the last clamp.

Remove excess glue from the panels right after taking off the clamps (usually within two hours of the initial clamping if you use aliphatic-resin glues and your shop is above 65°F). If you dally and let the glue become brittle, it will invariably take small, but deep, chunks of wood with it when you scrape it off.

Once the glue has reached full strength (the next day is ideal), surface

To secure a panel for surfacing with hand planes, fasten two back-to-back clamps to a piece of plywood, which is, in turn, screwed to the work platform. Begin surfacing with a No. 4 smooth plane set coarse and shoot at a diagonal across the panel (left). Then run a No. 4½ plane, set to take a finer cut, with the grain and parallel with the edge of the boards (below).

and smooth the panels. If there are a lot of them, consider taking them to a commercial shop that has an industrial thickness sander. It will not seem cheap, but the labor you save will (theoretically) allow you to do more work, and probably more enjoyable work, over the course of a year. And the results should leave the panels ready for final sanding. Before running the panels through the machine, remember to transfer the module code from the faces to the edges.

If you must surface the panels yourself, do so on a good work table. Position a pair of 32-in. lifts in the open space in the southern half of the shop, set up the 4x8 platform and screw a piece of plywood approximately 32 in. wide and 48 in. long to it. Clamp a pair of back-to-back clamps across the plywood to hold the panels securely for surfacing, as shown at right and above.

Depending on your energy level at the time, you can use planes and cabinet scrapers or a belt sander for this operation. Planing and scraping are very taxing, even when the tools are extremely sharp and their beds have been waxed to reduce friction. My feeling, however, is that it goes faster than belt sanding. I'm quite sure the surfaces end up flatter, and I know that it's a lot quieter.

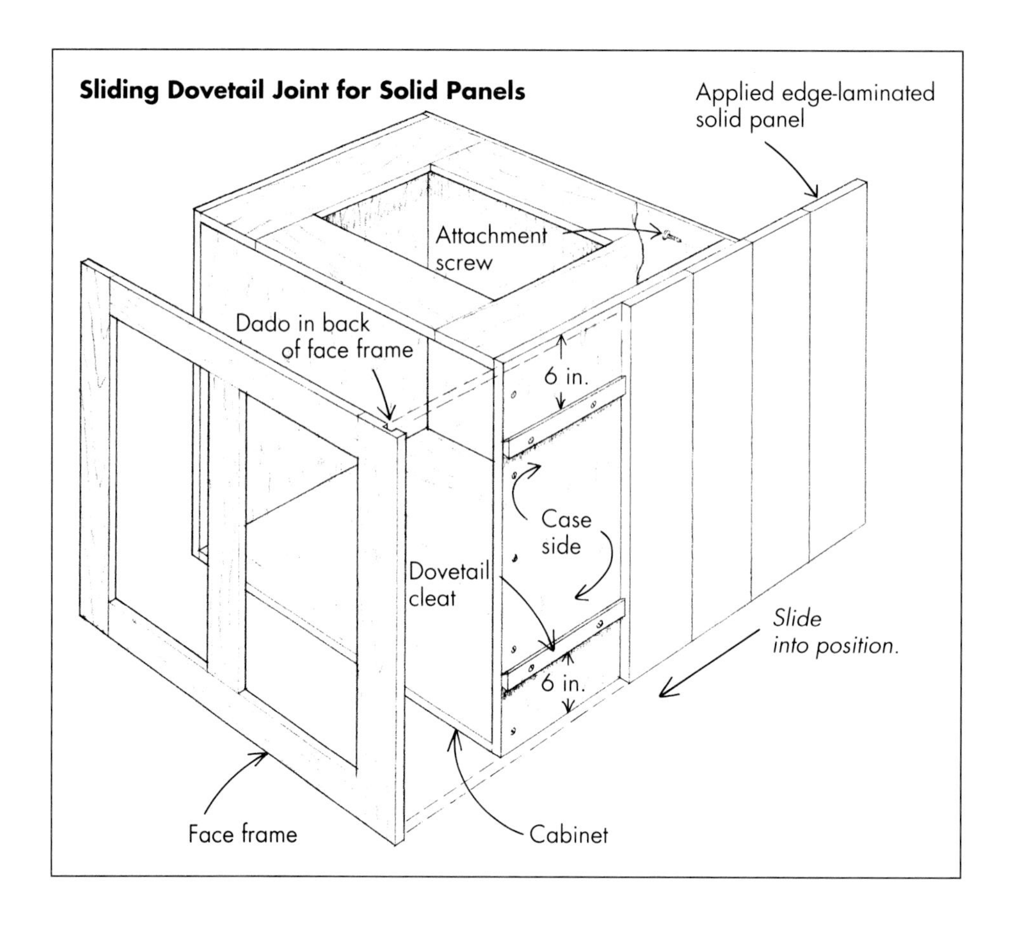

If you opt for planing, start with a number 3 or 4 smooth plane set to take a coarse shaving, then follow with a number 4½ smooth plane (a wider version of the number 4) set to take a fine shaving. Use an 800-grit waterstone to knock the corners off the plane blades to avoid grooving the surface.

Shoot the number 3 or 4 plane diagonally across the boards from the leading edge to the rear edge. Work your way down along the length of the panel and repeat the process until you've contacted every part of the surface. Then use the number 4½ plane to clean up the marks left by diagonal planing. Run this plane the full length of the panel, being sure to follow the grain of the wood. (You may have to reverse the direction of planing if interior boards have an opposing grain direction.) Keep the blade sharp to ease your labors. Use the cabinet scraper to clean up any defects. I find that scrapers mounted to an adjustable handle are the easiest and most effective to use. (Mine has a ball-and-socket joint and is available from Robert Kaune; see Appendix I on pp. 140-142.)

After scraping, the panel is ready for final sanding. Provided your scraper was sharp, you should be able to start in with 180-grit paper on an orbital finishing sander capable of holding half a sheet of sandpaper. I recommend Stikit self-sticking paper made by 3M—it lasts longer than many standard sandpapers and, best of all, it allows you to change the paper in less than 10 seconds. You must, however, replace the base pad on your sander with the inexpensive pad sold by the suppliers of the paper (see Appendix I on pp. 140-142).

Lightly dampen the surface of the panel and sand it again, using 220-grit paper. For a smoother finish, continue to 320 grit. Stack the surfaced panels with their frames along the east wall to the left of the drill press.

If you've decided to use the belt sander, start with 80 grit, running the sander back and forth along the length of the boards. Unless you have more than one sander, do all the panels at 80 grit before changing belts to 100 grit and then to 120 grit. Finish the job with the orbital sander. Depending on the manufacturer of the sandpaper, you can probably start at 120 grit and continue through 150, 180 and 220 grit.

The next step is to cut the panels to final width and length, and sort them by function: banked drawer faces, framed panels and solid applied panels. After cutting out individual drawer faces from the panel, re-mark the pyramid and module symbol on the back of each face and stack them with the rest of the drawer faces. It's likely that the framed panels and drawer faces will receive some form of edge treatment, so load them on Larry, wheel them to the area by the table saw and perform this operation with the router mounted under the side extension table. Stack the panels along the east wall of the shop. Collate panels that are to receive a frame with these components.

If the project calls for solid applied panels, they must be specially milled to allow for shrinkage and expansion once they are attached to a cabinet side; otherwise, the panels will split. I use a sliding dovetail joint for this, attaching a dovetail cleat to the case side and running the mating keyway across the panel back. The panel is attached with screws from inside the case to the rear of the panel and allowed to "float" on the front end. The panel rides on two cleats secured to the case side with drywall screws, as shown in the drawing on the facing page. The front edge is buried in a dado in the face frame and can float in and out inconspicuously. (If the cabinets are not going to receive a face frame, the front edge of applied panels will show. In this case, the dadoes must be stopped before they reach the front edge of the panel.)

Make the joint by first routing two straight-sided dadoes in the back of the applied panel one-half the depth of the stock and ½ in. wide. Locate the dadoes 6 in. from the top and bottom edges. Use the back-to-back clamp to guide the router base. Once the dado is cut, insert a dovetail bit, set it to the depth of the dado and run it along the dado's length to produce an undercut on one side. Move the clamp so that a second pass will produce an undercut on the opposite side.

After routing the keyways in all the applied panels, create the dovetail cleats. Transfer the bit from the hand-held router to the router at the shaping station. Select a piece of stock the thickness of the widest portion of the keyway and run it against the bit. (Adjust the fence and the height of the bit on a test piece until the cleat matches the keyway with just enough play to allow movement.) Store the completed panels and dovetail cleats by the carcase components along the south wall of the shop to await finishing and carcase assembly.

Preparing Drawer Faces

Now bring all the drawer faces to the 4x8 platform to drill for the adjustment cams, which attach the face to the box while allowing the face to be moved up to 3⁄16 in. in any direction. This feature allows the alignment of drawer faces to be fine-tuned with surrounding face components, such as doors, other drawer faces and trim elements. Make a jig (as shown in the drawing on p. 76) to locate the 20mm holes that must be bored into the back of the drawer faces. Two cams per drawer are sufficient, except for faces wider than 6 in. Note that the jig provides a choice of holes for locating two or four cams per face.

After locating the cams on the backs of all the drawer faces, bring the stack of

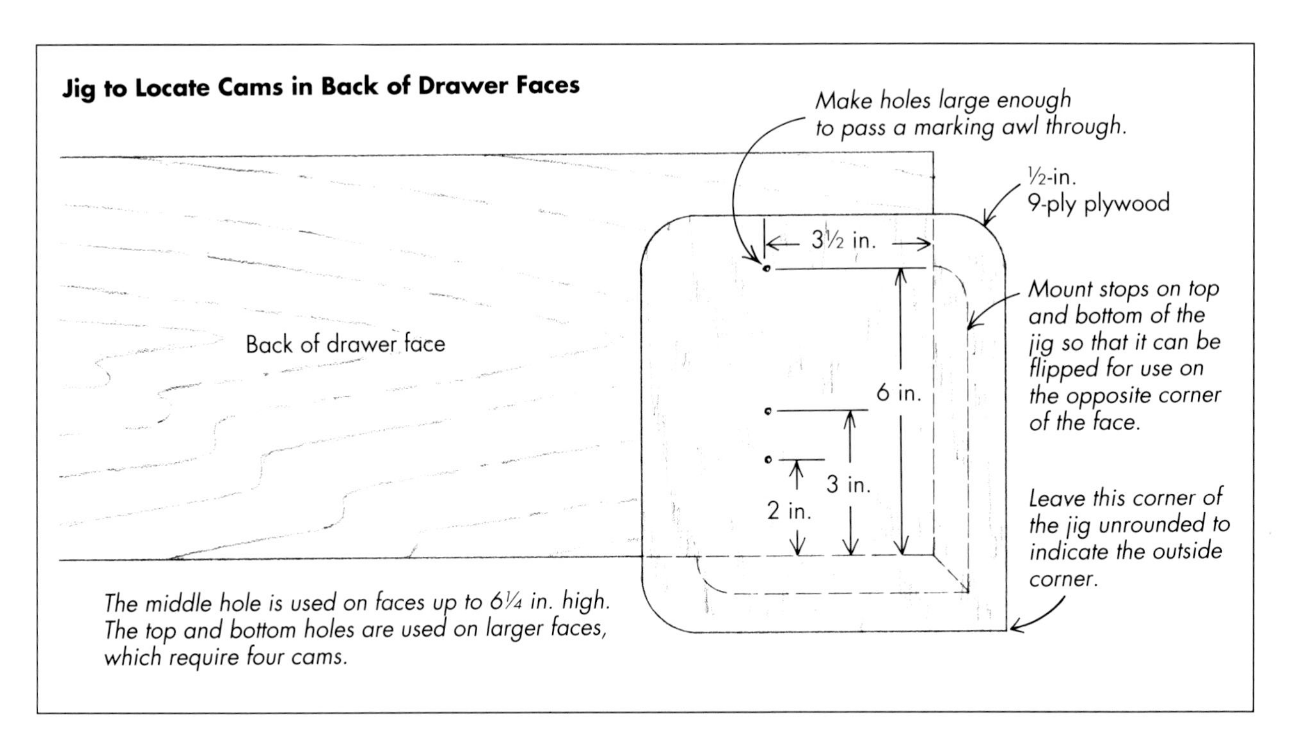

Holes are drilled into the back of the drawer faces to receive the adjustment cams that attach the faces to the drawer boxes. Here, steel centering pins are placed in the cam holes to locate the shank holes in the front of the drawer boxes. Note the piece of carpet laid on the work platform to protect the drawer faces.

faces over to the drill press. Chuck the 20mm bit into the machine and drill a test hole to fix the depth stop. (The cams should be flush to the surface when pounded all the way in.) Proceed to drill the cam holes, centering the brad point of the bit on the pinhole made by the awl. When the process is complete, wheel the stack over to the shaping station if a molded edge is needed.

Finish-sand the faces the same way you did the applied side panels. (If these panels were a part of the project, the setup is already in place.) Then stack the faces with the drawer-box components.

Preparing Other Solid-Stock Components

There are three remaining components to make up from the solid stock: the exposed nailers, the side and back supports for slide-out shelves and the carcase top frames. Optional components, such as appliance garages, and cubbyhole shelves below upper cabinets, are also processed at this time. The exposed nailers have already been cut to width and length, so locate them in the stacks against the east wall and load them up on Larry, the materials stooge, or Moe, the mobile tool stooge. Sand the wood to 180 grit and bring the completed nailers to the stack of carcase components along the south wall of the shop.

The design I use for slide-out shelves (see the photo and drawing at right) requires that two wood sides, a front and a back be joined to a bottom of ¾-in. thick hardwood plywood. Bring these components to the work area from the south wall. The joinery is accomplished with spline biscuits and finish nails. Standardize the procedure by using a stick marked with the slot centers for the spline biscuits. Use one face of the stick for laying out the sides and the other for the fronts and backs. Transfer the marks to the components and make the slots. Hold the material securely against the 4x8 platform, and clamp a 2x4 across the platform for a back stop. I like to round the front edges of the side supports and break all the edges with a 3⁄16-in. roundover bit. Then sand the stock to 220 grit. Collate all the components into their individual units and stack against the south wall.

Now bring the top frame components to the 4x8 platform and separate the sides from the fronts and backs. Prepare the stock for the spline biscuits using the jig designed for the splining of door frames (see the drawing on p. 71). Alternatively, set up on the stationary machine (see the photos on p. 70). When slotting is complete, re-sort the components into their individual units, but don't bother to stack them along a wall. You are now ready for the next major production block—component assembly—and the top frames might as well be the first project.

A slide-out shelf in operation. Note that the door is on hinges that carry it out of the way of the slide-out.

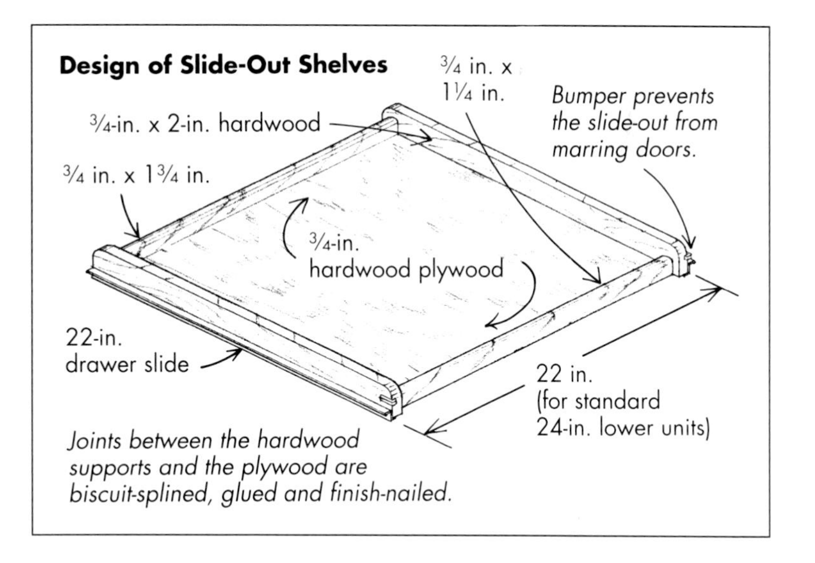

Chapter 11: Component Assembly

The component-assembly production block serves to clean up all the loose ends proliferating in the shop. Components to be assembled include carcase top frames, doors, panels, face frames, drawers and slide-out shelves. In addition, edge bands are applied to any sheet-stock components that require them. All the assembled components are sanded in preparation for finishing, thereby avoiding the need for disruptive sanding operations during that production block. After finishing, all that remains in the construction phase of the project is to assemble cases and install the components.

Carcase Top Frames

The top frames are already in place on the 4x8 platform, and you need only add spline biscuits and clamps to assemble them. Begin by laying out one unit in its proper configuration, then dry-fit the splines in place. If a joint doesn't close, check for a misoriented or shallow slot. If the fit looks good, apply glue to the walls of the slots, insert the splines and clamp the components together. (A glue applicator made by Lamello has a dual-vented tip that applies glue to both walls at once, making the process faster and less messy; see Appendix I on pp. 140-142 for the address.) Mark the time on the frame and continue assembling units until you're out of clamps.

When the frames are dry, remove the clamps, scrape off the excess glue and bring the adjoining surfaces flush with a block plane. These components are hidden from sight and touch, so sanding isn't necessary. Since each of the frame components has been cut ½ in. long, the assembled frames must be trimmed to final dimension. Use the sliding table (see the photo on p. 52) to square one edge, then use the rip fence for the next two cuts. First, cut the frame to width, then, with the frame turned 90°, cut it to length. Unless your rip blade cuts very coarsely, don't bother to rip the pieces 1⁄32 in. oversize and plane to final dimension. Stack the completed frames along the south wall of the shop to one side of the carcase components.

Panels and Frames

Bring all the panels and frame bundles up to the platform from their holding area to the left of the drill press. Untape a set of frames and find the corresponding panel or panels. Set up the pipe-clamp-support fixture (if it's not up already), dry-fit the panel and continue filling the fixture with dry-fit units until you run out of room. Glue up each unit, taking great care to keep the joints flush along the edges and the panel perfectly square. Use a 12-in. square against the inside edge of the frame (as shown in the photo on p. 80) and if the clamps allow enough clearance, check the diagonals for squareness. If the panels are solid wood, rub paraffin from a candle on the corners to prevent the glue from cementing the panels to the frame. (Unless solid panels are allowed to float in their frames, the wood is likely to split during periods of shrinkage.)

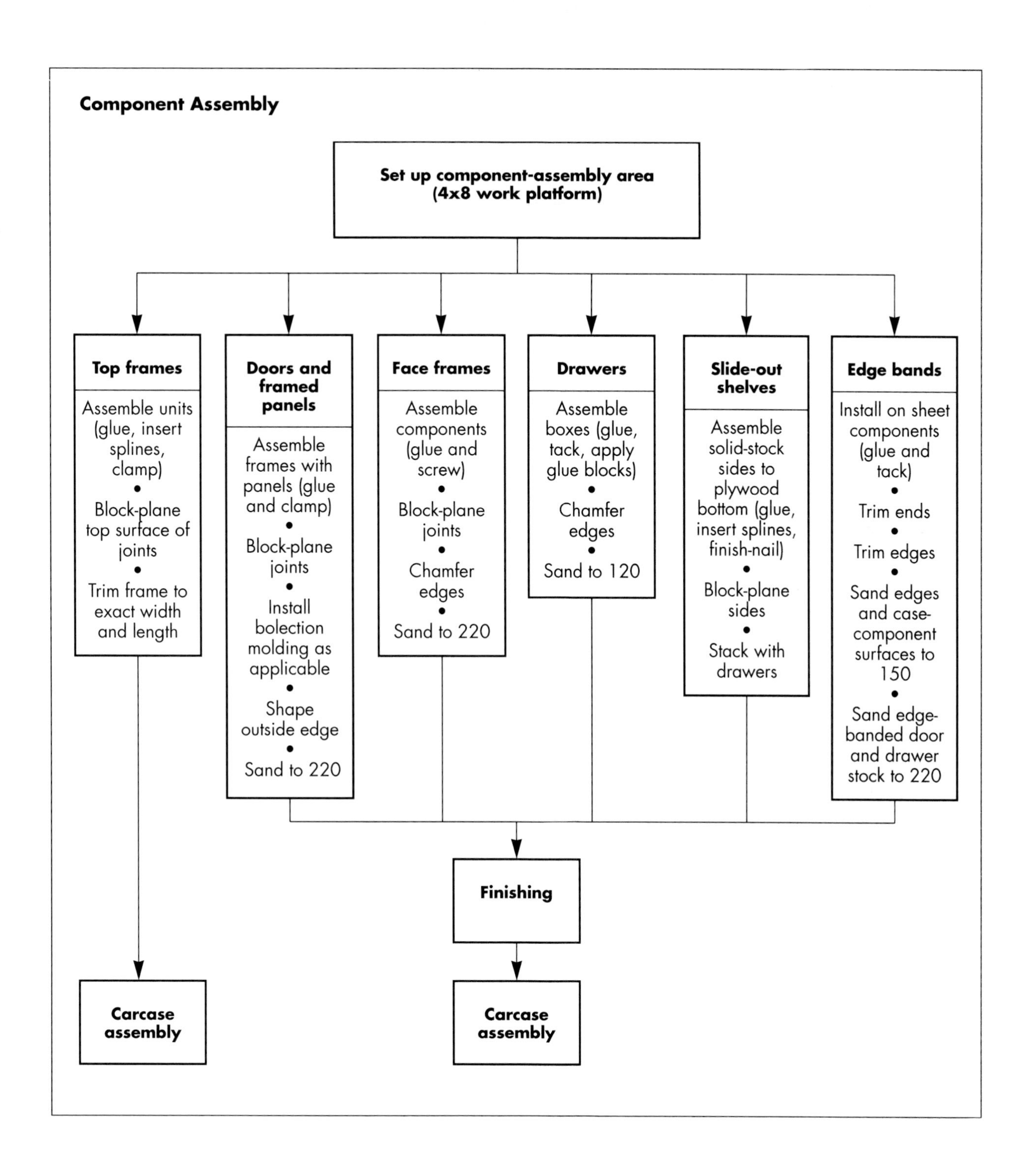
Component Assembly
Set up component-assembly area (4x8 work platform)
Top frames
Assemble units (glue, insert splines, clamp)
•
Block-plane top surface of joints
•
Trim frame to exact width and length
Doors and framed panels
Assemble frames with panels (glue and clamp)
•
Block-plane joints
•
Install bolection molding as applicable
•
Shape outside edge
•
Sand to 220
Face frames
Assemble components (glue and screw)
•
Block-plane joints
•
Chamfer edges
•
Sand to 220
Drawers
Assemble boxes (glue, tack, apply glue blocks)
•
Chamfer edges
•
Sand to 120
Slide-out shelves
Assemble solid-stock sides to plywood bottom (glue, insert splines, finish-nail)
•
Block-plane sides
•
Stack with drawers
Edge bands
Install on sheet components (glue and tack)
•
Trim ends
•
Trim edges
•
Sand edges and case-component surfaces to 150
•
Sand edge-banded door and drawer stock to 220
Finishing
Carcase assembly
Carcase assembly

Immediately after gluing up a frame-and-panel unit, check it for square by holding a small steel square against the inside edge of the rails and stiles. Make adjustments by lightly tapping the unit with a rubber mallet.

Remove the clamps as soon as possible (remember to chalk the time of clamping on the units), reclamp a new unit and begin cleanup by scraping off the excess glue. Clean up the joints between rails and stiles with a block plane and cabinet scraper. After all units have been glued up, released and cleaned up, wheel them to the shaping station to receive their molded edge treatment. (If, however, you're only going to break the sharp edge with a block plane or an ⅛-in. roundover bit in a laminate trimmer, do this on the work platform.) Begin sanding with 150-grit paper on a ¼-sheet orbital sander. Moisten the wood slightly to raise the grain at the 180-grit stage and final-sand at 220 grit. Stack the completed units against the wall to the left of the drill press to await finishing and hardware installation.

Face Frames

If there are face frames in the project, stack the taped bundles on Larry, the materials stooge (see p. 32), and roll them to the 4x8 platform. Open one bundle and lay out its components in the correct configuration. Since the face frame will be screwed together from the back, the components must lie face down on the table. It's tricky at first to picture the layout of a face frame from the perspective of the *inside* of the cabinet, but you soon get used to it. If you get confused, face the elevation card away from you and hold it up to the light. You are now looking at the face frame from inside the cabinet.

Once the components are properly oriented, mark the position of the interior joints. This must be done to careful measurement, unless the components themselves can be used as spacers. (For example, the location of a mid-stile that carries a mid-rail can be determined by sliding the mid-rail first to the top, and then to the bottom of the mid-stile, thereby establishing its position to the fixed distance of the mid-rail, as shown in the drawing on the facing page.)

Begin fastening with the innermost joints of the frame. Clamp the component to receive the screws along the edge of the 4x8 platform, using two C-clamps at the location of the joint, as shown in the photo on the facing page. Butt the component to be fastened firmly against the clamped piece at its marks and drill pilot holes for the face-frame screws. Back off the free piece, clear out the chips and apply glue to its end. Reposition the piece and screw the fasteners home. If other components are to be joined to the clamped piece, repeat the process. Now release the clamps, lift up the joined components and remove the excess glue with a damp rag. Wipe off the work surface, too. Continue until all the frame components have been joined, then check with diagonal measurements to ensure the frame is square. (You may have to clamp a temporary brace across the structure.) Set the assembly aside to dry. Continue until all the units have been assembled, and let the entire batch sit overnight to allow the glue to reach maximum strength.

Clean up these frames the same way as any others: Block-plane the surfaces and edges of the joints and sand the wood to 220 grit. I break the edges of the wood inside the frame and out with an ⅛-in. roundover bit in a laminate trimmer. Set the completed frames to the right of the drill press to await finishing and hinge-plate installation.

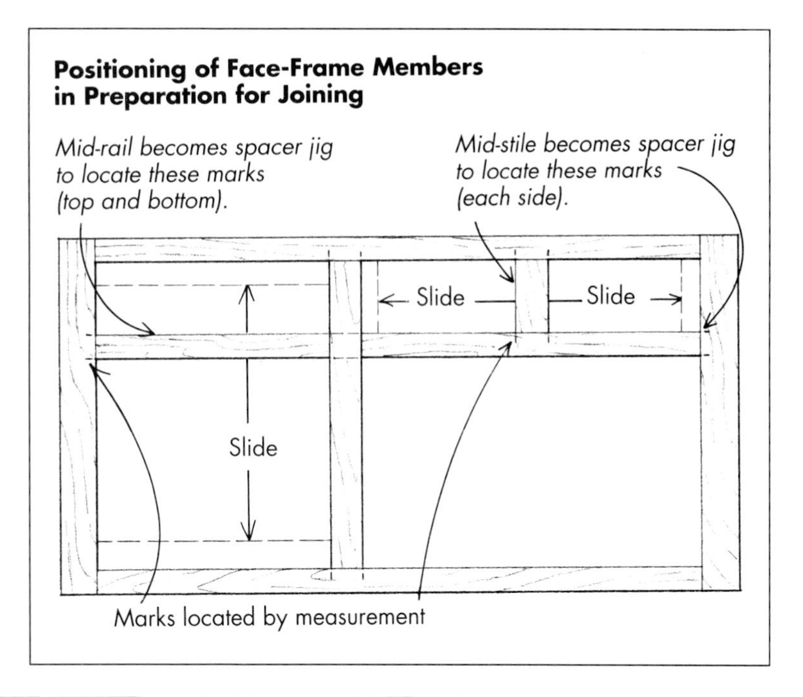

When assembling a face frame, clamp the piece to receive the screws securely to the work surface. The tendency for the pieces to slide out of flush is great, so it's a good idea to use two C-clamps at each juncture.

To assemble the drawer boxes, wrap the sides and fronts around the ¼-in. bottom panel and check to make sure that the joints fully close at each corner (right). Then glue and tack the joints together (below).

Glue blocks support the bottom and hold the boxes square. Use ⅝-in. slight-headed brads to hold the blocks in place while the glue dries.

Drawers

The first step in drawer assembly is to sort the pieces (excluding the faces) into units. The sizes of the drawers are listed on the elevation cards; transfer the module code to the undersurface of the drawer bottom and stack the units on the work platform, leaving enough room to assemble one unit.

Since the box depends on the squareness of the bottom, double-check this piece with diagonal measurements before assembling each unit. Take care of slight discrepancies with a block plane. When satisfied, surround each bottom with its box, checking to be sure that the joints close completely (see the top photo on the facing page). Apply glue lightly to both surfaces of the joint and tack the structure together with ⅝-in. slight-headed brads, as shown in the bottom photo on the facing page. Set the box upside down, check for square and then glue ¼-in. by ¼-in. by 2-in. softwood glue blocks along the juncture of the bottom with the sides, two to a side, to prevent the boxes from racking (see the photo above). Now bring the box over to the area where the feed platform was formerly set up. Place it on a piece of plywood raised up on 12-in. lifts and allow the glue to dry thoroughly.

Continue to bring finished boxes over to the drying area, carefully stacking them upside-down, one on top of the other. Wait until the next day to move the boxes back to the platform, then break the edges of the plywood with the ⅛-in. roundover bit. Sand the boxes inside and out to 120 grit. Return them to the drying area to await finishing and installation of hardware and faces.

Slide-Out Shelves

Bring the components for the slide-out shelves to the 4x8 platform, sorting the solid stock and ¾-in. thick plywood bottoms into individual units. Write the module code on the underside of the bottoms. Begin assembly by inserting the spline biscuits and then shooting 6d finish nails with your air gun to hold the front and back to the bottom while the glue sets. Make sure the front and back are flush to the edge of the bottom on each end. If they are not, bring them flush with a block plane. Install the sides in the same manner. Stockpile the completed units with the drawer boxes.

Edge Bands

Carcase sides, floors, ceilings, partitions and shelves are banded along one edge only. Doors and drawer faces made of sheet stock are edge-banded along all four edges. Because of the enormous variation in lengths and the sheer number of pieces, I find it easiest to crosscut the bands at the time of assembly—an exception to the guiding principle of the block production method. Simplify the task by counting all the bands to be cut to similar lengths—all the carcase sides, for example—and precutting this stock. Be sure to add ¼ in. for final trimming when cutting edge bands.

Starting with carcase components, identify the edge to be covered (on case components, the orientation symbol appears on the back edge). Roll glue on the edge of the sheet stock and lay on the band. Tack it in place with slight-headed brads, making sure the band overhangs the edge all around. If you are not getting a tight joint between the edge band and the sheet stock, you will have to clamp the band, using a strip of ¾-in. by ¾-in. stock between the clamp heads and the band to distribute the pressure. Continue banding until all the carcase components (including the shelf stock) are done. If doors and drawer faces are to be edge-banded, apply banding to two opposite edges. Trim as described below, and apply the two remaining edges.

Allow the glue to dry for several hours at room temperature, then scrape off the excess glue and begin trimming. I use the table-saw method introduced by Paul Levine in his book on cabinetmaking (see Appendix II on p. 143) to trim overhanging ends. Remove overhanging edges on the router table using a piloted trimming bit, as shown in the drawing on the facing page. An alternative method, which is easier to use on larger components, is to mount the same bit in a hand-held router. Make sure to keep the router bed flat on the edge band. A special arrangement of back-to-back clamps holds the panels upright on their edges, as shown in the photo on the facing page. Sand all the case components and shelving stock to 150 grit. Look before you sand, as there is no need to sand surfaces that will never be seen, such as case sides covered by applied panels and the bottom surface of floors. Stack these panels against the south wall to await finishing.

Edge-banded door and drawer faces are held out for edge treatment. This usually entails the breaking of edges with an ⅛-in. roundover bit mounted in a laminate trimmer. Sand the outside surfaces of these components to 220 grit. Store completed components against the wall to the left of the drill press. Door hinges are installed after finishing. Adjustable cams will be pressed into the backs of the drawer faces, and pulls will also be installed.

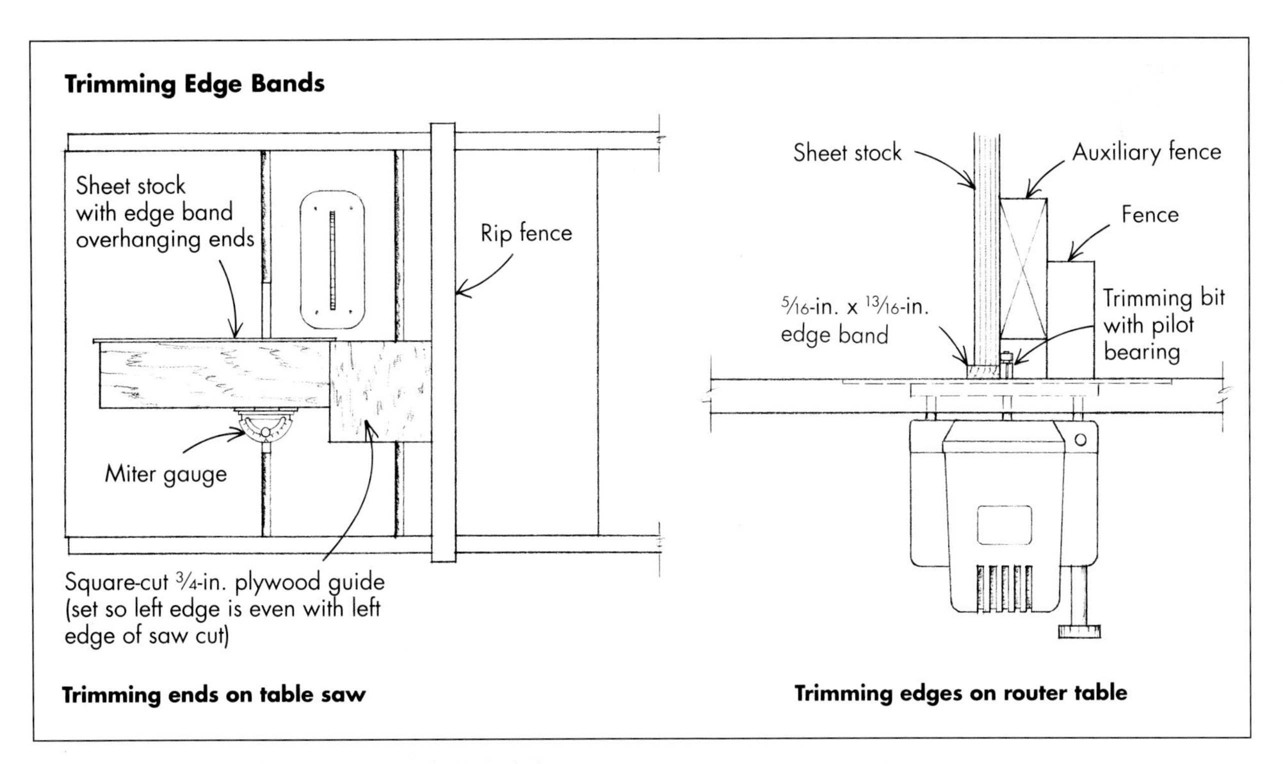

Back-to-back clamps hold a large case component upright while the edge band is being trimmed. Note that pipe clamps (without their heads) have been secured vertically to the lifts to support the bottom of the panel. The board to which the two back-to-back clamps have been secured is temporarily screwed to the end of the work platform.

Chapter 12: Finishing

My cabinetmaking colleagues tell me that I take a rather unique approach to scheduling the finishing of my products, for unlike most cabinet shops, I finish all the components before final assembly. I remind these people that my methods of building the cabinets are just as unorthodox, as is the size and overhead of my shop. I work the way I do in keeping with one of the basic rules I introduced during my discussion of the small-shop work style: "No project will be assembled until there is a place outside the shop for it to go." I don't have room for my cabinetwork when it is put together, so the last step in the production process has to be the case assembly. After that, the job is out the door.

Finishing Materials

In just the past few years I have made a major shift in the type of finish I apply to my cabinetry, swinging from the penetrating oil finishes, most recently tung-oil varnishes, to the new, high-quality water-based lacquers. I'd wanted to use lacquers for a long time, because of their great resistance to abrasion, heat and water. They dry fast, too—the thought of applying a finish to an entire set of kitchen cabinets and delivering them the same day was almost too much to bear. Having to provide a spray room in which to use the lacquer, however, was also too much to bear.

Then I discovered water-based lacquers, in particular the ones made by Hydrocote (for their address, see Appendix I on pp. 140-142). They have all the characteristics of the nitrocellulose-based lacquers and a number of advantages: They build up faster and dry harder, and because they are nontoxic and nonflammable, they don't require a spray booth. I finally had my lacquer, and could eat it too. (Actually, the manufacturer doesn't recommend eating the stuff, even if it is nontoxic.) Although I still occasionally have a client who insists on an oil-and-varnish finish (the depth and rich luster of a penetrating finish is, admittedly, unobtainable with a surface finish), today I use lacquer for most of my work.

Water-based lacquers tend to spray a little differently from their nitrocellulose counterparts, but I had no trouble learning to use them. Hydrocote offers a videotape, other instructional materials and workshops to teach proper application. More and more information is also becoming available in books and in the periodical literature.

Preparation

Whether using lacquers or penetrating oils, there are certain ways in which you should prepare the wood, and the shop, for finishing. First and foremost, remember that the quality of the finish directly depends on the quality of the surface preparation of the wood. This means sanding, and the finer and more thorough the sanding, the higher the luster of the finish. Try to follow the suggestions for sanding mentioned in the previous chapters, and don't neglect to raise the grain of the wood at least once at

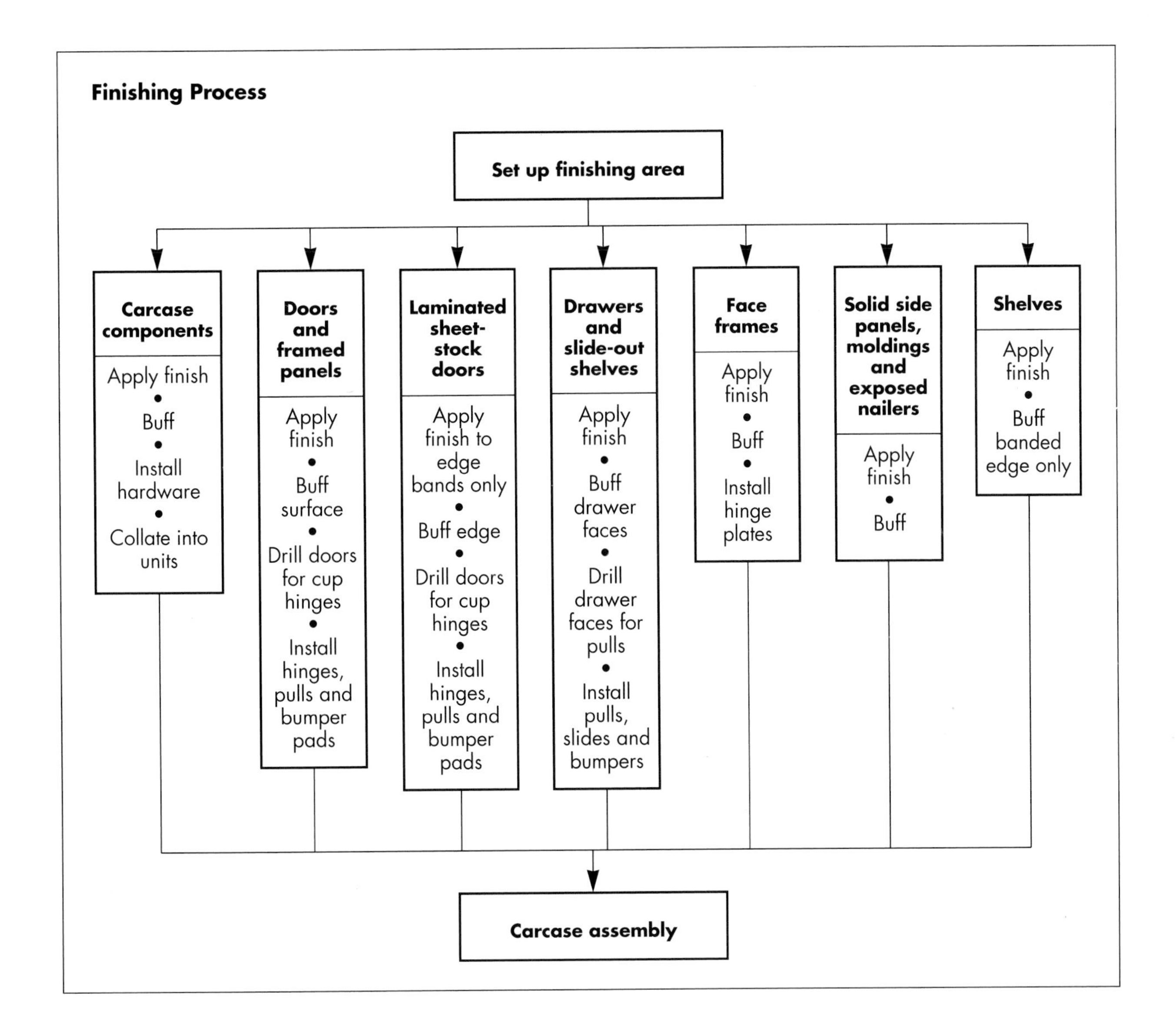

the 180-grit stage. (The first spray of a water-based lacquer or sealer will raise the grain again somewhat; this is taken care of by sanding between coats with 220 grit.) When sanding is completed, vacuum the shop thoroughly and run the dust collector or an exhaust fan to remove the dust particles from the air. But don't dampen the floor to keep down the dust if you are going to spray water-based lacquer: It tends to blush (turn white) under high humidity. Make sure the temperature of the shop is above 65°F before applying the finish. (Water-based lacquers love the dry heat from woodstoves, and, unlike nitrocellulose lacquers, they can be applied in the vicinity of an open flame.) The lacquer should also be above 65°F, so don't store it on the floor of the shop during the cold months, especially if the floor is a concrete slab.

Set up the finishing area in the southern half of the shop by placing two 8-ft. long 2x4s across a pair of 32-in. lifts. Secure the 2x4s to the lifts with drywall screws and run a self-sticking padding (weatherstripping works well) along the tops of the 2x4s to protect the surfaces of the components. Set blocking on the floor by the west wall to receive the components during intermediate stages of the finishing process.

Finishing Procedures

The basic finishing procedure for all components, regardless of the type of finish used, entails the following steps:

- Check the surface for scratches and remove them with a cabinet scraper, then sand with 220-grit wet/dry paper.
- Apply the finish to all accessible surfaces and set the piece aside to dry along the south wall.
- After drying, reposition the piece to apply finish to the surface missed in the previous step.
- Sand the surface lightly with 220 grit to remove dust bumps and raised grain.
- Reapply a second coat of finish to all surfaces.
- Rub out the second coat with 0000 steel wool and apply a final coat. (If you are finishing with water-based lacquer, use 320-grit wet/dry paper instead of steel wool.)
- Buff out the finish on all components except interior panels with rubbing compound on a wool pad. Buff either by hand or, more effectively, with an electric drill or grinder equipped with a buffing pad.

If the wood is a porous species such as oak, ash or cherry, you should use a sanding sealer for the first lacquer coat. This seals the pores and allows better buildup of the surface film. Porous woods drink up oil finishes, leaving almost no residual oil on the surface after the first coating, and sometimes even after the second coating. (Residual oil must be wiped off the surface after about 15 to 30 minutes, before it becomes tacky.)

Preparation for Case Assembly

Once the finishing process is complete, prepare the components to be assembled into cases. Beginning with the carcase components, install the sockets for the adjustable legs onto the floors and screw the drawer slides and hinge plates into the appropriate holes on partitions and side panels. (Large-shank "system screws" are used to attach this hardware to the predrilled 5mm holes.) Before restacking the components against the south wall, collate them into their modules. Include the vinyl-coated back, the top frame (if the module is a lower unit), nailers (upper units and freestanding hutches) and other unit-specific case components, such as spacers and braces for kitchen accessories (lazy Susans, roll-out baskets, etc.). In each unit, stacking the vinyl cabinet backs first allows them to act as spacers between the modules.

Now drill the backs of the doors for the cup hinges. If you have a drilling and insertion machine for your brand of hinges, simply set the right- and left-hand stops to coordinate with the spacing of the hinge plates on the case sides and start drilling (see the top photo on the facing page). The drawing on p. 90 shows the details of laying out this spacing. The machine also presses the hinges into place, as shown in the bottom photo on the facing page. (Note that the hinges can still be removed by backing out two screws; the plastic inserts to receive them remain pressed into the 5mm holes.) The only preparation remaining for the doors is to install the pulls and rubber bumper pads. Use a jig (see the photo on p. 91) to locate the shank holes for the pulls to speed up the process and to eliminate alignment problems.

You can, of course, drill and install European cup hinges without special machinery. Simply drill a 35mm hole to receive the hinge, using a fence and homemade flip-down stops similar to the ones on the mini-press (shown in the photos on the facing page) to hold

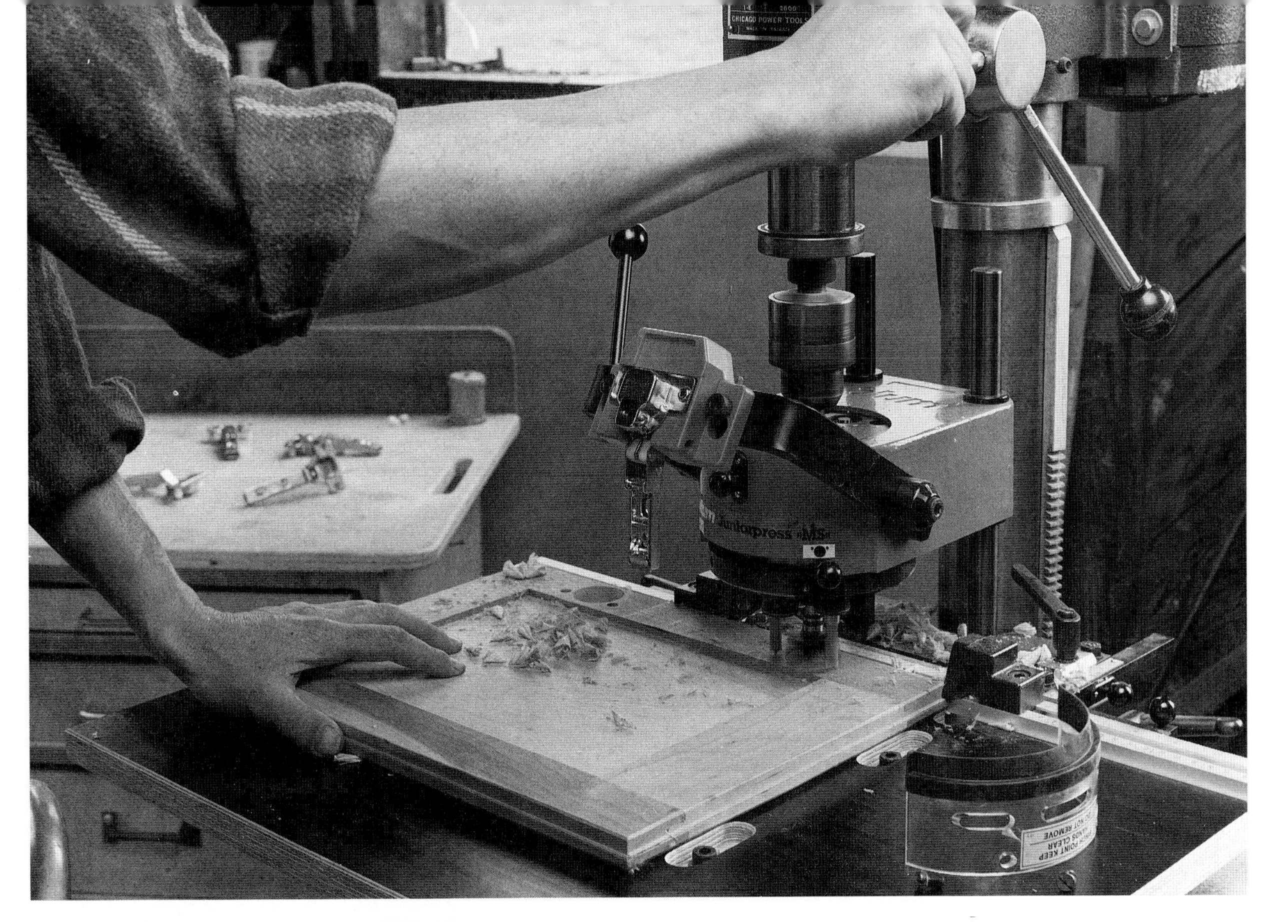

A hinge-boring machine attached to the drill press will bore all three holes for the European cup hinge into the back of a door at the same time (above). An accurate stop and fence system holds the door in the correct position (guard removed for clarity). The hinge-boring machine also presses the hinge into place (left). An arm holds the hinge out of the way while the drilling occurs and is then swung into place for the insertion process.

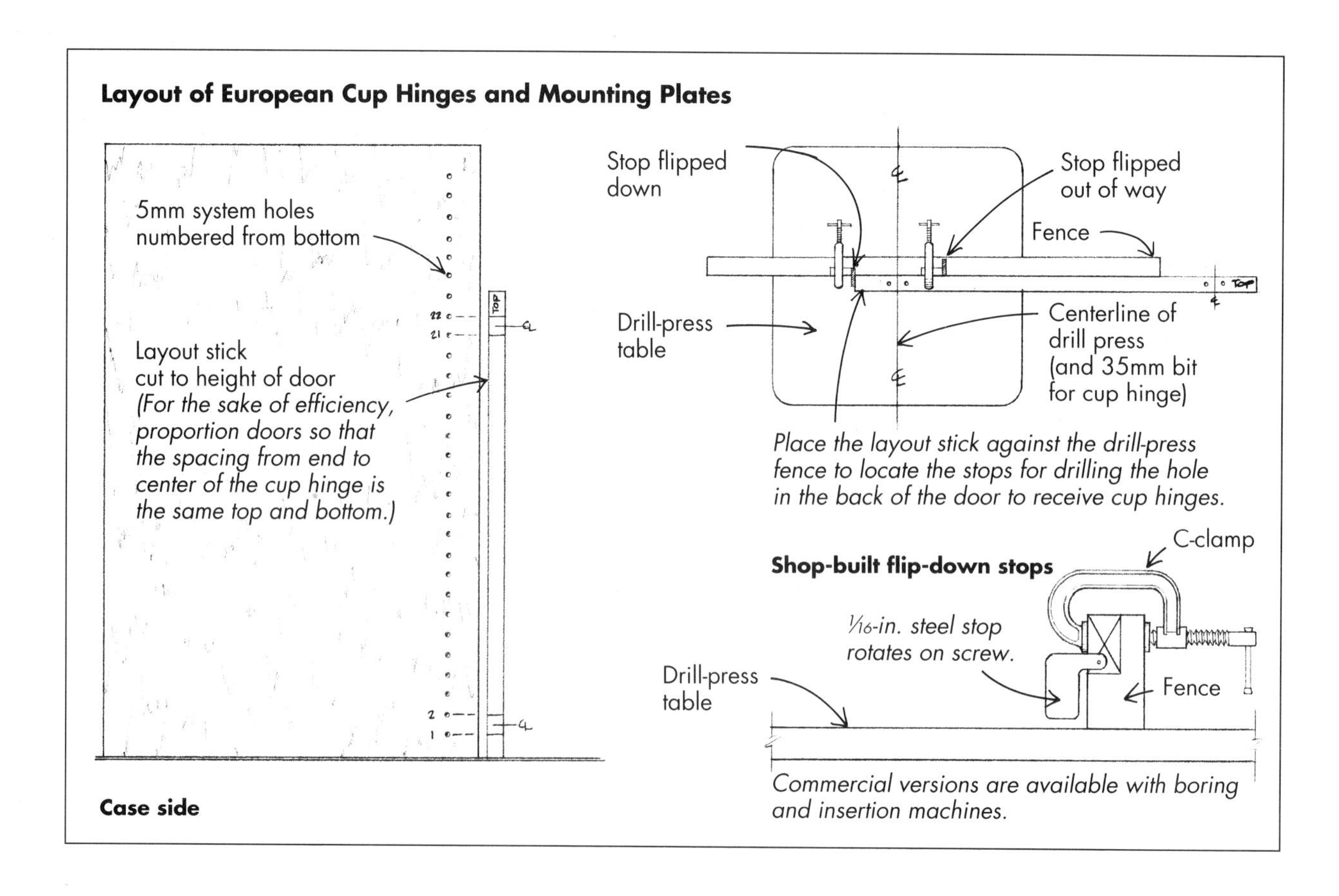

the door in place. Then press in the hinge by hand. Align the fastening plate parallel to the edge of the door and drill pilot holes for the attachment screws, using a self-centering bit. (I recommend Vix bits; see Appendix I on pp. 140-142 for suppliers.) Once the hinges, pulls and bumpers are installed, return the doors to the left of the drill press.

The other components that require hardware are the drawer faces, drawer boxes, slide-out shelves and face frames. Prepare for them by removing the 2x4s from the 32-in. lifts and setting the 4x8 platform in its place.

Begin with the drawer faces. Put them face up on the platform on a piece of soft cloth and drill shank holes for the pull hardware using a jig, as shown in the photo on the facing page. You can devise your own jig or purchase a commercial model (see Appendix I on pp. 140-142 for suppliers). Install the pull and set the faces down in the area with the drawer boxes.

Now bring the boxes to the 4x8 platform and install the drawer slides. Be sure the front edge of the slide is snug against the face before fastening it in place. (The Vix bit is helpful in locating and drilling pilot holes for the fasteners.) Also install the drawer slides for the sliding shelves at this time, and attach the door bumpers to each front corner. Return completed drawers and sliders to the north portion of the shop.

The final components to receive hardware are the face frames, if the project requires them. Bring each frame over to the work platform and, you'll be relieved to hear, place it right side up. Install the mounting plates for the cup hinges along the edges of the frame in the appropriate location. I recommend using a commercial jig for this purpose, because it will have to stand up to

A jig made by the Häfele company locates the proper position for the shank holes of a variety of pull styles. Here, a bar-type pull requiring two holes 76mm apart is being prepared.

a lot of drilling (see Appendix I on pp. 140-142 for suppliers). Be aware that hinge plates come in a variety of sizes, depending on how far the door overlaps the frame. Hinge plates for different overhangs look very much alike; make sure you are installing the right ones by trying out a door after the first pair has been installed. Relocate the completed frames to the right of the drill press.

Chapter 13: Carcase Assembly

The first step in the carcase-assembly process is to set up a level, flat platform on which to build the cases. This is easy to do—simply lower the 4x8 platform from the 32-in. lifts onto the 12-in. lifts. Insert cedar shim stock between the lifts and the floor to level the surface and prevent it from rocking.

The carcase components sit between the assembly platform and the south wall, sorted into modules. Bring the first unit onto the platform and begin assembly by securing the sides to the floor and top frame (or ceiling, if an upper cabinet). Assemble the case face up, so you can be sure all the components are flush on this plane. Use spline biscuits and drywall screws, or Confirmat connectors, to join the pieces. If using the latter, hold the pieces in alignment with a specially designed clamp while the jigged step drill prepares the pilot and shank holes, as shown in the photo at left. (Both clamp and jig are available from Häfele; see Appendix I on pp. 140-142.)

Now install partitions, nailers, braces and spacers. Except for partition bottoms, which may receive spline biscuits (unless this is a knockdown assembly), self-tapping 1⅝-in. drywall screws or Confirmat screws are all that are needed to fasten these components into place.

Break the edges of the edge bands (assuming the project doesn't have face frames) with an ⅛-in. roundover bit in the laminate trimmer.

Now turn the case over and install the back (see the photo on p. 94). Align the back with the outside edge of the floor and staple it in place or secure with 1-in. drywall screws. Check the squareness of the case with diagonals and, when satisfied, secure an upper corner to hold the case square while driving home the rest of the fasteners. (I usually cut the back ⅛ in. smaller than the overall dimensions of the case to prevent it from interfering when taking diagonal measurements.)

A specially designed jig attached to the front of a hand drill guides a drill bit into the corner of a case assembly to create a pilot hole for a Confirmat connector. The drill bit is stepped and tapered to conform to the shape of the fastener. Confirmat fittings are extremely strong and can be drawn in and out of their holes many times without losing fastening power.

If the cabinets have face frames, perform the following steps: First, check the face of the box for squareness with diagonal measurements, and adjust accordingly. Second, retrieve the appropriate face frame from the stack to the right of the drill press. Third, roll glue onto the face edges of the case components and set the frame in place. (Check to be sure the position is correct by referring to the module's elevation card, which indicates the amount the face frame overhangs the side components.) Finally, secure the face frame to the case with 6d finish nails or countersunk screws (which will be plugged with wood bungs).

Installing the Other Components

Next come the rest of the parts—side panels, drawers, slide-out shelves, special hardware, moldings and doors.

The side panels may be installed in either of two ways, depending on the type of panel. If they are of solid wood, connect them to the case with the sliding dovetail described on p. 75. Align the cleat on the case by sliding it into the keyway on the panel; apply a tiny bead of glue to the cleat where it will contact the case side. (Rub paraffin on the adjacent panel surfaces to prevent them from sticking to the case side in the event of squeeze-out.) Position the panel on the case and hold it in place with padded guitar clamps until the glue dries. Note that the panel should fit into a dado cut into the back of an overhanging face frame. If there is no frame, it should protrude past the carcase side by the thickness of the door and drawer-face stock.

When the glue has set, slide the panel off and additionally secure the cleat to the case side with countersunk 1-in. drywall screws. Slide the panel back in place and fix it to the cabinet along its back edge by running in 1¼-in. drywall screws from inside the case. Don't sink the screws, but cap them with cover caps designed for this purpose (see Ap-

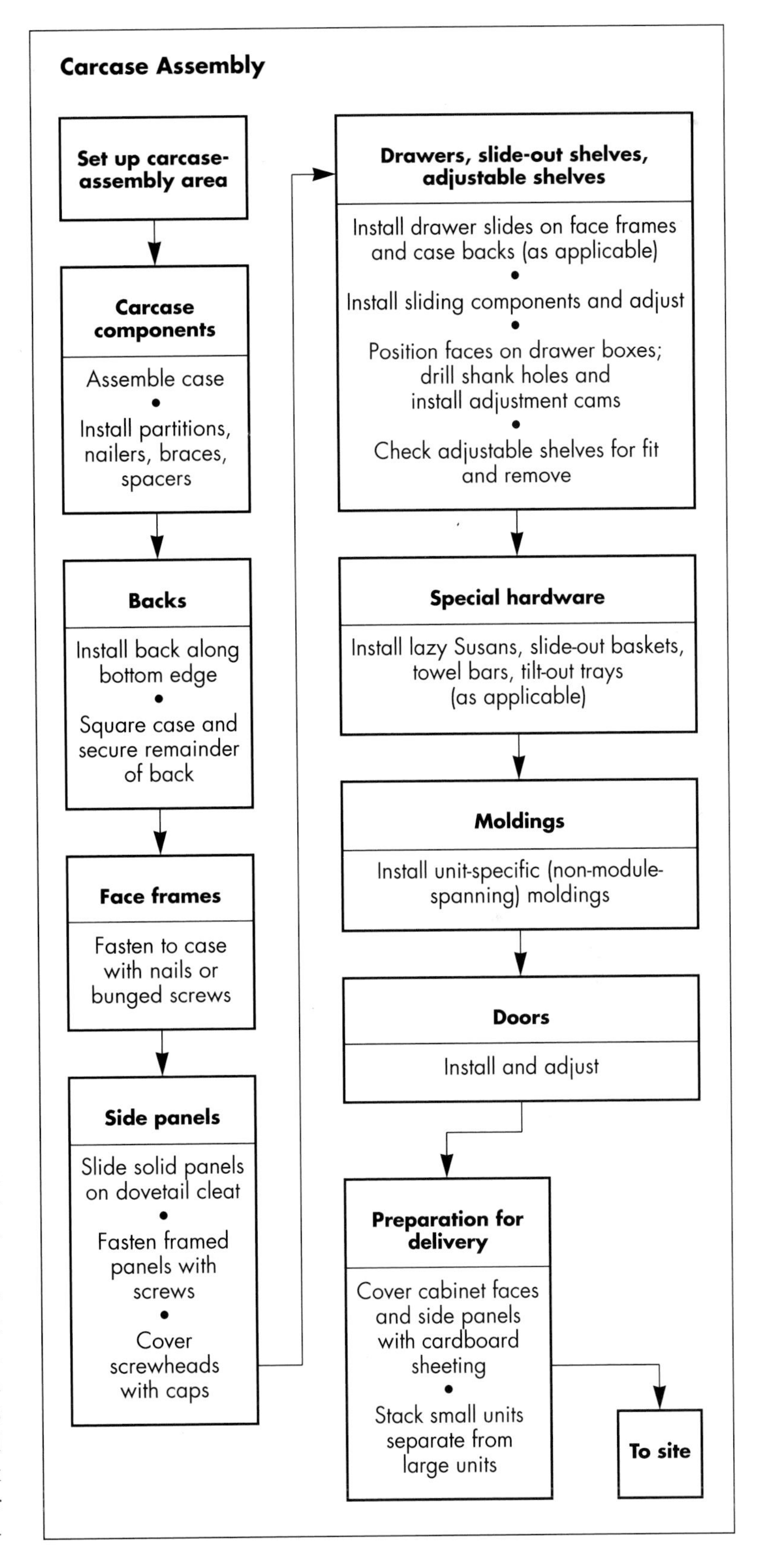

When fastening a back to a case, first attach it along only one edge. Confirm that the box is square by checking diagonal measurements, then tack a corner to hold the box square while the back is screwed into place.

pendix I on pp. 140-142 for suppliers). Any movement in the panel will now take place in the position of the front edge, which is either buried in a dado or standing free of the case. The back edge, which will be scribed to the wall during installation, will not move and will thus maintain a perfect fit.

Compared to the previous process, attaching framed panels is simple. Because a framed panel is relatively stable dimensionally, it's not necessary to provide for its movement. Just position the piece and secure it from inside the case. Be sure the screws are attaching the frame and not the panels, which must be free to move within the frame. Cover the screwheads with caps.

If the cabinet has a face frame, install the drawer slides now. Use the jig made by the manufacturer of your slides to hold each slide in position while you secure it to the side of the face frame and to the case, as shown in the photo on the facing page. If a side or partition is flush to the inside edge of the frame, simply screw the slide to the side or partition. Otherwise, a rear-mounting socket catches the back of the slide and is attached to the case back with screws (use only two screws initially, one in the horizontal adjustment slot and one in the vertical slot).

With the drawer slides in place (in faceless cabinets, the slides are installed immediately after the case sides are finished, as discussed on p. 90), install the drawer boxes. The faces are now attached to the boxes as follows: Insert metal centering pins in the 20mm holes drilled for the adjustment cams (available from the supplier of the cams; see Appendix I on pp. 140-142) and hold the face in its intended position on the cabinet. Press the face against the drawer box so that the pins mark the position of the shank holes. Remove the face and drill the shank holes in the front board of the drawer box. Remove the centering pins and tap the adjustment cams into place. (Protect the face during this step with a piece of carpet on the work platform.) Attach the drawer face to the box. When closed, the drawer face should touch the face frame equally on all four sides. If not, adjust the position of the slides either by moving the rear-mounting brackets along the adjustment slots, or moving the rear of the slide up and down along the slotted hole in the slide. Occasionally you may need to shim the slide out; strips of fine sandpaper are ideal for this. When a good fit is achieved, secure the brackets by screwing through the holes provided.

Install slides for the sliding shelves in the same way as the drawer slides. Slide the unit into place to check for smooth operation and then remove it (for shipping). Also check the fit of the adjustable shelves in the case. Find the shelves for the unit—they should be marked on the back edge—and position them. They shouldn't bind along the sides as they travel up and down.

At this time, also install any special fixtures, such as lazy Susans, slide-out baskets, towel bars or tilt-out trays. The larger items often require special braces in the case to receive mounting hardware. Cut these to fit and install with drywall screws. After installation, check all the fixtures for proper operation.

Most moldings are applied to the cabinets after they have been installed, because they generally span more than one module. If, however, the unit has moldings that are integral to the piece, cut these to fit and install with finish nails. Backing nailers (strips of wood) will have to be provided for certain moldings, such as side spacers and cornices. I fill the nail holes with a color-matched putty. Sometimes a standard color will be close enough; otherwise I mix various colored putties until I create a good match.

The doors are the last components to be installed in the cases. If all has gone according to plan, the hinge arms will slide right into the mounting plates without a whimper. I'm always amazed when it happens, as it does 99 times out of 100. All that remains to do is to adjust the hinges until an equal margin of space appears around the perimeter of the door. The ingenuity of these hinges becomes apparent as you realize you are given three planes of adjustment to work with: in and out, side to side and up and down. If the door abuts a drawer face, remember that the face is also adjustable on its inset cams. It does not take long to equalize all the margins. Of course, the slight racking and settling of the carcases during installation will require that you do just a little more adjusting then.

Your final task on the doors is to check that the proper hinge has been used. On certain cabinets (for example, those with roll-out shelves), the door hinge must be a wide-angle type. This variety of hinge carries the door completely clear of the opening so that slide-outs are unobstructed when extended. In other cases, such as when two doors share a partition, the overhang is half the normal amount. It's far better to discover an error in hinge selection in the shop than miles away at the job site.

A plastic jig manufactured by the makers of the drawer slides (in this case, Blum) holds the slide in perfect alignment while the rear-mounting socket is secured. This method of attaching slides is necessary only in cabinets with face frames.

Preparation for Delivery

Before taking the completed cabinet off the assembly platform, prepare it for an event-free ride to its new home. I buy 4x8 sheets of corrugated cardboard from a local paper-supply outlet and cut sections of it to cover the faces and finished side panels of the cabinet. The doors and drawers are left installed (except in large units where their weight becomes a factor) and are thus secured in place. Fix the cardboard to the cabinet with packing tape applied with a handled roller.

Now lift the mummified cabinet off the platform and put it near the door along the west wall of the shop. As cabinets continue to be assembled and stacked, keep the larger ones (which will be loaded first) separate from the smaller ones. If the project is large and promises to overwhelm the shop, try to deliver the cabinets to the site as you go along. An alternative approach is to construct the cabinets with knockdown fittings, assembling them on site. But I recommend that you avoid face-frame cabinets in this situation, as they have a substantially greater number of components and hardware parts to deal with than frameless cabinets.

Chapter 14: Installation

As the time for installation draws near, the specter of disaster invariably rears its ugly head. What if the cabinets don't fit the space? What happens if the moldings don't resolve in that tricky corner? What if the client doesn't like the work? You may never have to look this specter in the eye if you have laid down the foundation of a successful project: Consultations with the client were thorough, measurements and layout parameters were carefully checked and double-checked and the cabinets were designed to account for variables in the wall, floor and ceiling surfaces to which they would be attached. In the delivery of the cabinets, careful packing ensures that they arrive free of profit-rending damage, and consultation of a thorough checklist ensures that everything needed for installation is with you on the site.

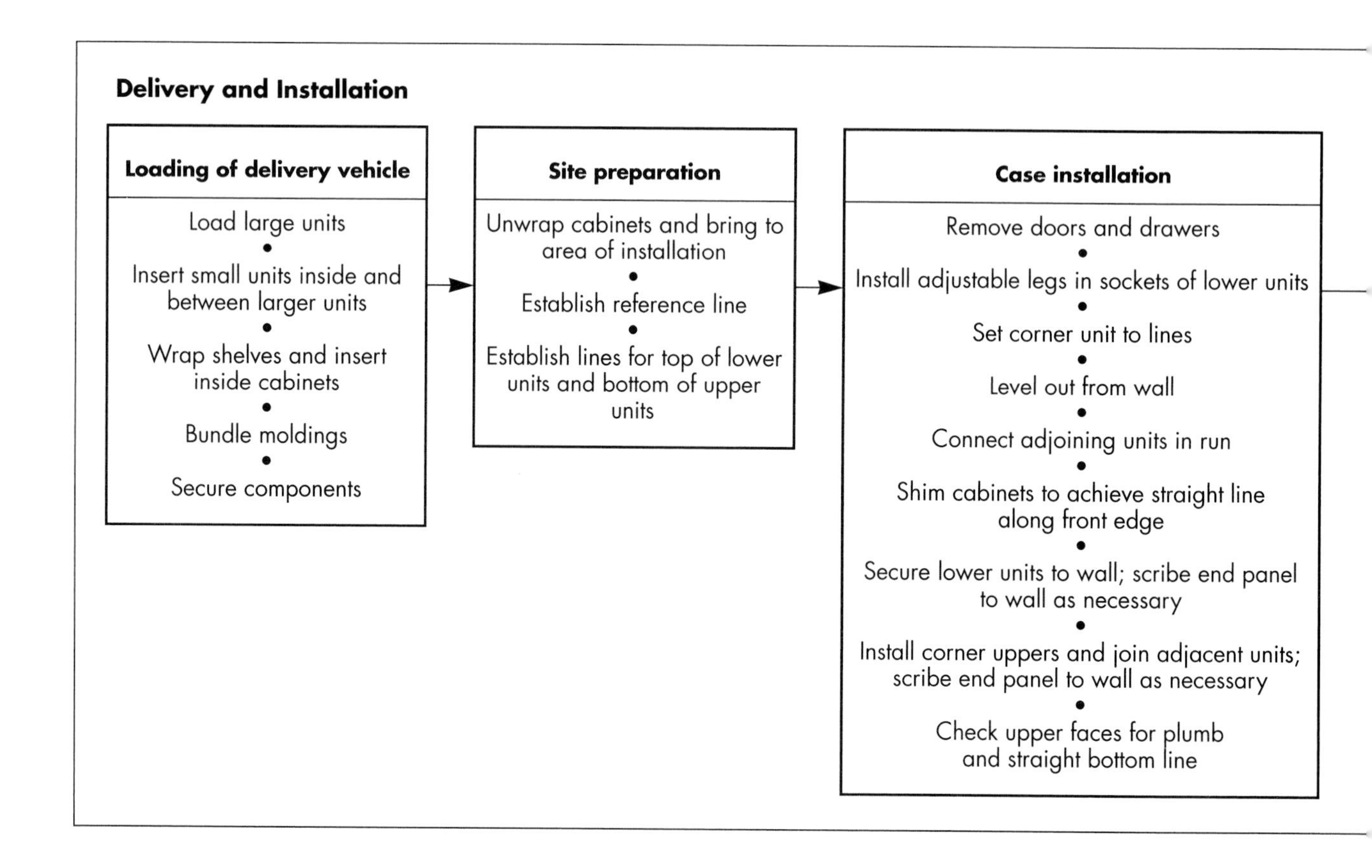

The Delivery

When loading the cabinets for delivery, put the larger units on the truck first. Smaller units can then be placed around and (occasionally) inside them. The larger units can also contain cardboard-wrapped bundles of adjustable shelves and other loose items. Moldings should also be bundled and taped securely together. Face the good surfaces in so that only the backs of the moldings are exposed. Pack the cases tightly so components will not jostle around during the ride; fill gaps with chunks of cardboard or moving blankets.

Whether you are delivering the cabinets in your own truck or a rented truck, be sure the insurance covers the contents of the vehicle. It is possible to add a one-time rider to your policy to cover the replacement cost of the cabinetry involved in a specific delivery. If you have hired a trucking company, read the details of its coverage and be sure to retain your copy of the freight contract and bill of lading, which specifies what you have placed on the truck.

Preparing for Installation

Over the years I have performed scores of installations, and I have yet to make it to the end of one without discovering the need for a tool or piece of hardware that remained behind in the shop. There always seems to be some new wrinkle that demands something I don't have with me. Fortunately (depending on how you look at it), there are usually hordes of other tradesmen on the site who can offer the missing item. I have, however, developed an effective checklist (shown on p. 98) that itemizes all the tools, materials and hardware that you should take with you to a typical installation; check off the items as you pack them into your truck.

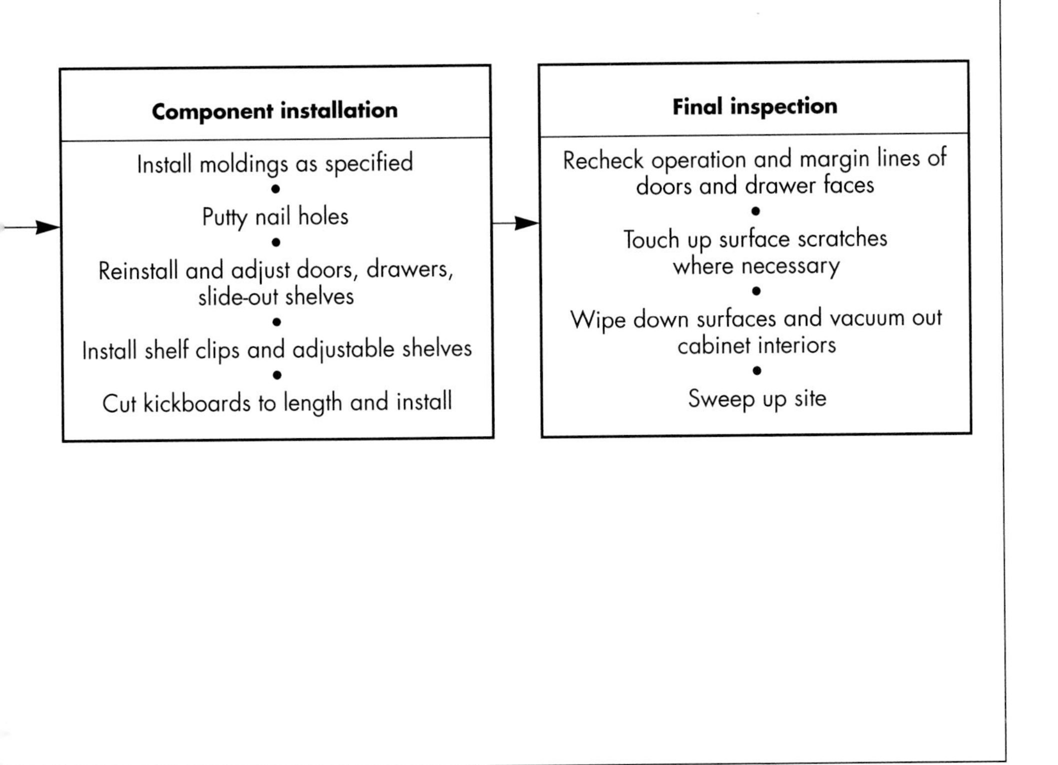

Installation Checklist

Tools

- ☐ Hand truck
- ☐ Air compressor and guns
- ☐ Chopsaw and stand
- ☐ Circular saw w/extra blades
- ☐ Compass saw
- ☐ Double-edge Japanese saw (Ryobi)
- ☐ Hacksaw
- ☐ Jigsaw w/extra blades
- ☐ Cordless drills and chargers
- ☐ Drills, drivers, holesaws for above
- ☐ Router w/bit selection
- ☐ Block plane w/extra blades
- ☐ 13-oz. hammer
- ☐ Tack hammer
- ☐ Nail sets and punch
- ☐ Set of chisels (¼-in.to 1½-in.)
- ☐ Tape measure
- ☐ Combination screwdriver
- ☐ Pliers
- ☐ Crescent wrench
- ☐ 24-in. square
- ☐ 12-in. combination square
- ☐ Bevel gauge
- ☐ Torpedo level
- ☐ 30-in. level
- ☐ 78-in. level
- ☐ Water level
- ☐ Plumb bob
- ☐ String with pin hooks
- ☐ Clamps (C- and deep-throat)
- ☐ Scissor jacks and support box
- ☐ Utility knife
- ☐ Pencil scribe
- ☐ Pencils and sharpener
- ☐ Stud finder
- ☐ Work apron
- ☐ Caulk gun
- ☐ Spotlights
- ☐ Extension cords w/junction box
- ☐ Sawhorses w/carpenter's vise
- ☐ Vacuum w/attachments
- ☐ 2-ft. stepladder
- ☐ 5-ft. stepladder

Miscellaneous

- ☐ First-aid kit
- ☐ Dust mask and earplugs
- ☐ Soap and paper towels
- ☐ Dust covers
- ☐ Cotton rags
- ☐ Flashlight
- ☐ Business cards
- ☐ Final bill and job card
- ☐ Plans
- ☐ Calculator
- ☐ Scale rule
- ☐ Drinking water
- ☐ Lunch
- ☐ Radio

Hardware

- ☐ Adjustable legs (two extra)
- ☐ Kickboard mounting clips
- ☐ Shelf clips
- ☐ Door bumpers
- ☐ Screw covers
- ☐ Extra hinges w/mounting plates
- ☐ Extra pulls
- ☐ Uninstalled hardware fixtures
- ☐ Drywall-screw selection
- ☐ Brad and finish-nail selection
- ☐ Oval-head #10 screws for uppers
- ☐ Lag screws
- ☐ Extra hardware screws

Materials

- ☐ Shim stock
- ☐ Colored filler putty
- ☐ Wood plugs for countersinks
- ☐ Masking tape
- ☐ Latex caulk (color-matched)
- ☐ Epoxy
- ☐ Yellow glue
- ☐ 220- to 320-grit sandpaper
- ☐ Touchup finish material
- ☐ 0000 steel wool
- ☐ Solvent
- ☐ Rubbing compound
- ☐ Molding stock w/extra lengths
- ☐ Scrap plywood

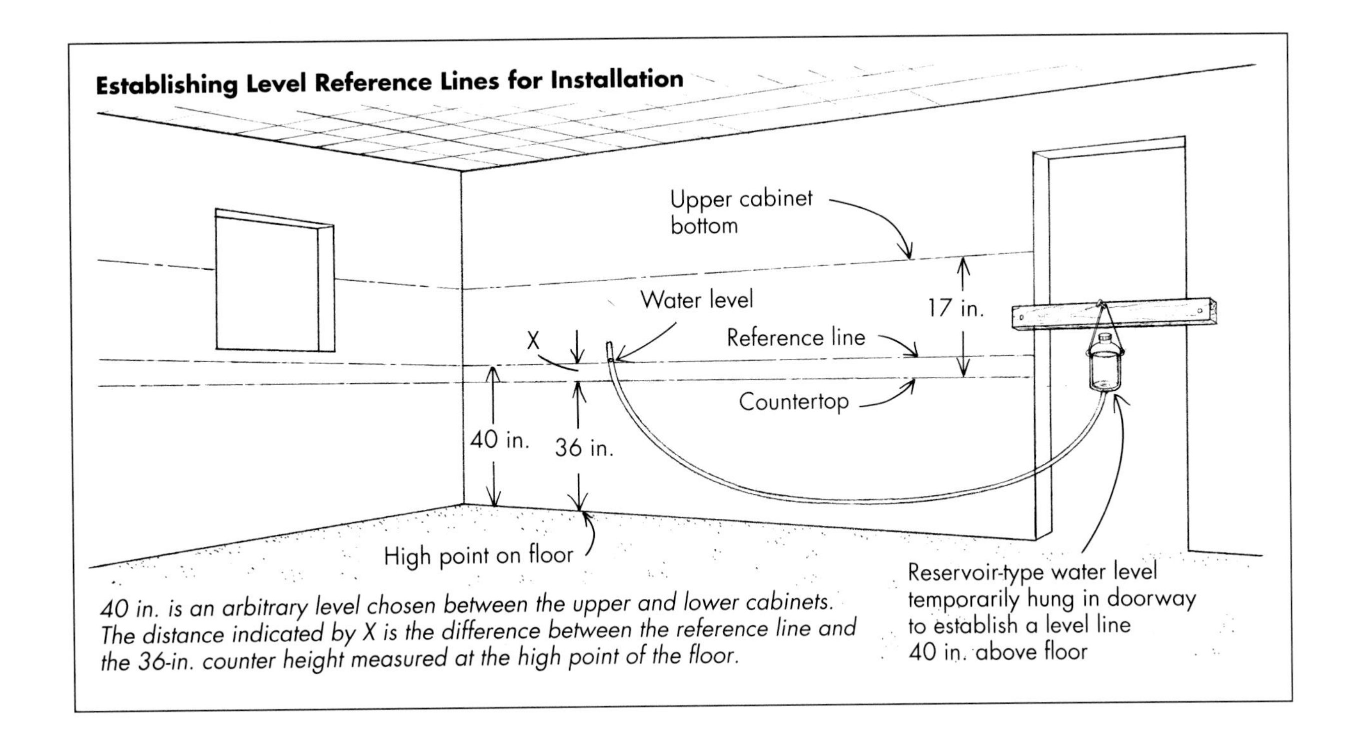

40 in. is an arbitrary level chosen between the upper and lower cabinets. The distance indicated by X is the difference between the reference line and the 36-in. counter height measured at the high point of the floor.

Once you're at the job site and have brought all the cabinets, tools and materials into the building, strip off the cardboard wrappings and check for damage. Unless the cabinet has to go back to the shop for repair (usually, the worst damage will be scratches and minor impact dings, which can be corrected on site), bring it into the general vicinity of its final home. When all the cabinets have been inspected and distributed, you are ready to begin installation.

Reference Lines

Unfortunately, you can never depend on a floor, wall or ceiling to be plumb, level or square to an adjacent surface, even in new construction—I once saw a floor deviate from level 1½ in. over the span of 8 ft. Thus the first task during installation is to establish a reference line for each run of cabinets.

There are several ways to do this. The fastest and most accurate is to use a reservoir-type water level, as shown in the drawing above. (I use the one made by Price Brothers Tools; see Appendix I on pp. 140-142. It's spillproof and may be operated by one person; nonreservoir water levels need two operators.) Without a water level, you will have to establish the reference line with a 78-in. or 96-in. level. Check that the level is accurate by setting it on a flat surface, shimming up one end until the center bubble reads level and then flipping the level over and taking another reading. The bubble should again read level; if not, loosen the bubble's setscrews and adjust until you can flip the level without producing a deviation. Adjust the other horizontal bubbles to match the corrected one.

The reference line should be established somewhere between the bottom of the upper cabinets and the counter surface (40 in. is a standard height). Draw the line by connecting marks established with the water level or by using the long level as a straightedge. Now measure down from the reference line to the floor at a number of places until you find the high spot of the floor. Standard practice is to set the counter

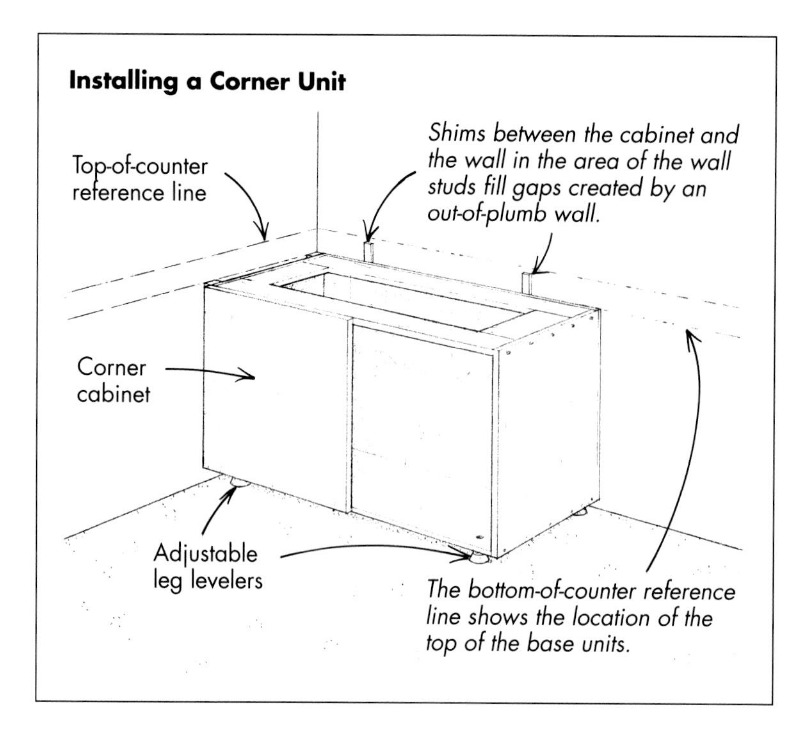

surface 36 in. above this point. Assuming the reference line is exactly 40 in. above the floor at this point, a mark made at 36 in. will fall exactly 4 in. shy of the reference line (distance X in the drawing on p. 99). Establish the counter-height line around the entire kitchen by making marks the distance of X, in this case 4 in., below the reference line and connecting them with a straightedge. Assuming a spacing of 17 in. between the counter surface and the bottom of the upper cabinets, mark this distance above the counter line everywhere there will be upper cabinets.

Installing the Cabinets

On the subject of whether to install the upper or lower units first, cabinetmakers seem to be divided into two camps. The first camp insists on installing uppers first because the lowers "only get in the way." The second camp would never install uppers without the lowers in place to run their jacks or braces to. It's really a matter of the way you intend to deal with the upper units. Without the lower units, you either need a strong, patient helper to hold the upper units in place while you jockey them into position, or a set of specially engineered cabinet jacks (see Appendix I on pp. 140-142 for a supplier). I am of the second camp, primarily because I happened to have a set of old scissor jacks when I installed my first set of kitchen cabinets.

I ready the lower cabinets for installation by removing the drawers and doors, and by snapping the adjustable legs into their sockets. Preset the height of the feet to 4 in., the theoretical standard for producing a 36-in. high counter with a 31¼-in. case and ¾ in. of countertop material. The first cabinet to be installed is usually a corner unit (see the drawing at left), as it is much easier to work your way out of a corner than into one—contrary to everywhere else in life. The corner unit acts as a fixed point from which the other cabinets begin their runs.

Begin installing the corner unit by setting the cabinet in its approximate position and then bringing the top even with the counter-surface level line (minus the thickness of the counter material). You'll appreciate the ease with which a simple turning of the adjustable legs allows you to do this.

If the corner cabinet is a full-height unit, you will not, of course, have a top corresponding to the counter-surface reference line. In this case, make a mark on the side of the unit at 31¼ in. (or the height of your lower units, if different) and use this mark to level the unit with the counter-height reference line (again, minus counter thickness). Plumb the face of the cabinet and attach it to the wall with screws through the nailer in the interior of the cabinet.

Assuming the corner cabinet is a lower unit, lay a 30-in. level across the cabinet top perpendicular to the wall and check for level, adjusting the front legs as necessary. Butt up the adjoining cabinets, level them to the reference line and perpendicular to the wall and then

While the applied panel is held plumb and slightly away from the wall, a pencil scribe transfers the shape of the wall to the edge of the panel for trimming. A piece of masking tape has been applied to the panel to make the line more visible—a standard procedure, especially with darker-colored panels.

attach them along the front edge, making sure the faces are flush. Use drywall screws if there are spacers between the cabinets, or specially designed connector screws if there are no spacers (see Appendix I on pp. 140-142 for a supplier). Install adjoining cabinets in the run in the same way. When all the lower units are leveled and connected, check the top front edge of the entire run for straightness, either with a stretched string or a long straightedge. Produce a perfect line by adjusting the cabinets in and out from the wall with cedar shims forced between the cabinets and the walls at the location of studs. Fasten the cabinets to the wall by running 3-in. drywall screws at a low angle through the top frame into the wall studs. If the last cabinet has an applied panel, scribe it to fit the wall and attach it to the case, as shown in the photo above.

With all the lowers secured, the upper units can now be set in place. Use jacks resting on a scrap-plywood box to hold the uppers at the proper height and level, as shown in the photo at right. Common auto scissor jacks are ideal for this purpose. Starting with a corner, secure the uppers to the wall with 3-in. number-10 oval-head screws (I have my doubts about the shear strength of drywall screws) through the nailers in the cabinet. Be sure you are hitting a stud. If necessary, insert shims between the cabinet backs and the wall to bring the cabinet faces plumb. Connect adjoining upper units along the front edge—make sure their faces are flush, then fasten them to the wall. Scribe exposed end panels to the walls.

Once the lower units have been secured to the wall, a box made of scrap plywood is laid down for the jacks that will support the upper cabinets.

A spacer molding is installed using an air-driven finish nailer. Alternatively, the molding can be screwed into place.

Installing the Components

Start with the various moldings, which span the cabinet modules and tie the design together. All are cut to fit on site with a chopsaw and are installed with finish nails (see the photo at left). Depending on the design of the cabinetry, you may be installing pilasters over the spacers between the modules, a plinth along the base of the lower units, a rail onto a nailer between the drawers and doors, a light pelmet under the upper units and a cornice along the top to tie the cabinets to the ceiling. Fill nail holes with matching putty.

Now reinstall the doors, drawers and slide-out shelves. Make final adjustments to the hardware to align and even out all the margins. Insert shelf clips into the 32mm-system holes and set the adjustable shelves in place.

Kickboards are the last pieces to be installed. As with the moldings, cut them to fit with the chopsaw. Use butt joints where they adjoin or intersect. Only the exposed end of a facing board is mitered (and returned to itself), as shown in the drawing below. Attach the kickboards to the adjustable legs with clips designed for this purpose. Since the clips fit anywhere along the groove

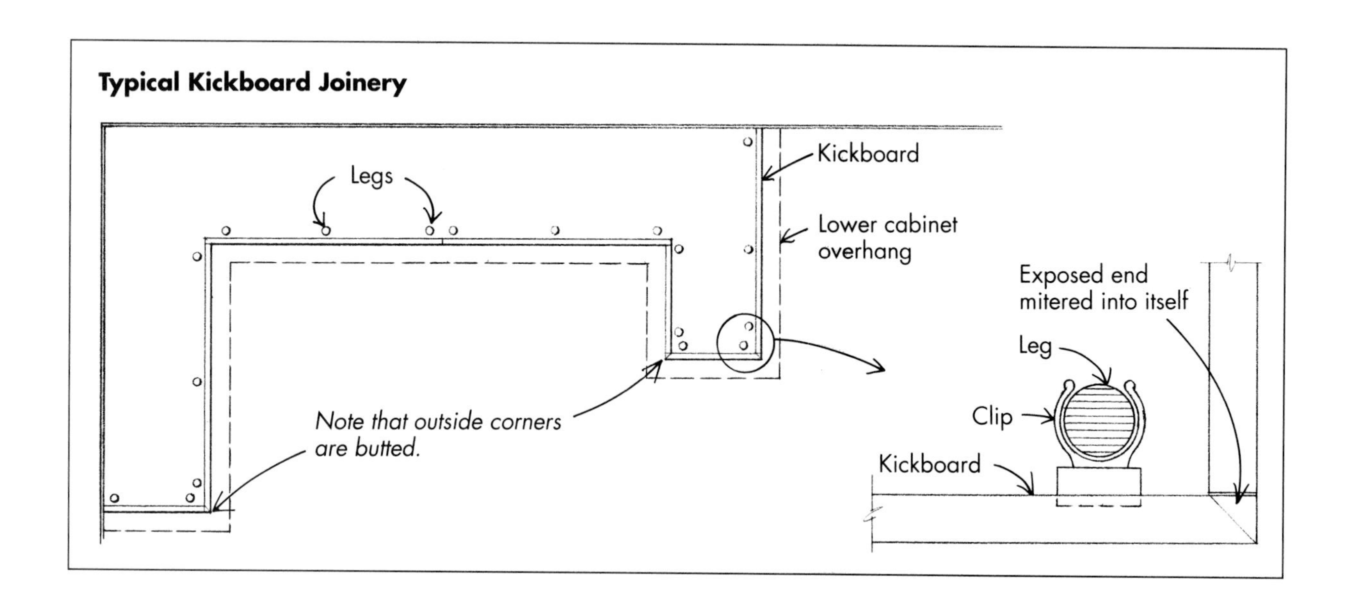

Kickboards are cut to length on site and then snapped into place to the leg supports.

provided for them along the kickboard's length, you can hold the board in its approximate position and mark the center of the legs. The clips are then pressed into place (the clips also slide on their mounts, so perfect alignment is unnecessary). The kickboard is then snapped into place (see the photo above).

Final Inspection

Installation is not complete until you have given the cabinets a thorough inspection. Check the operation of all the doors, drawers and sliding shelves. There should be no binding or sagging, and the margins should be as equal as you can get them. Be sure none of the bumpers are missing. If you come across a door that is so badly warped that hinge adjustments aren't enough to hold the door against the face equally at the top and bottom, you might try adding another hinge to add more force to the closing action. If this doesn't help, you will have to build another door. I always cut enough extra door stock to cover such a contingency.

Other things to look for include scratches in the finish and unputtied nail holes. Fill the latter, and sand the former with 320-grit wet/dry paper or 0000 steel wool; touch up with the finish material (either oil applied with a rag, or lacquer spread with a small brush). Complete the inspection by running a soft rag over all exposed surfaces to remove dirt and hand prints. Finally, vacuum out the interiors of all the cabinets, including the drawers.

After picking up all the scrap wood, candy wrappers and empty coffee cups, and loading all your tools and surplus materials into your truck, go back into the house and sweep out the rooms where you were working. The only thing you should be leaving behind is a beautiful set of professionally installed cabinets—and maybe a personal touch or two, as discussed in the next section.

Section III

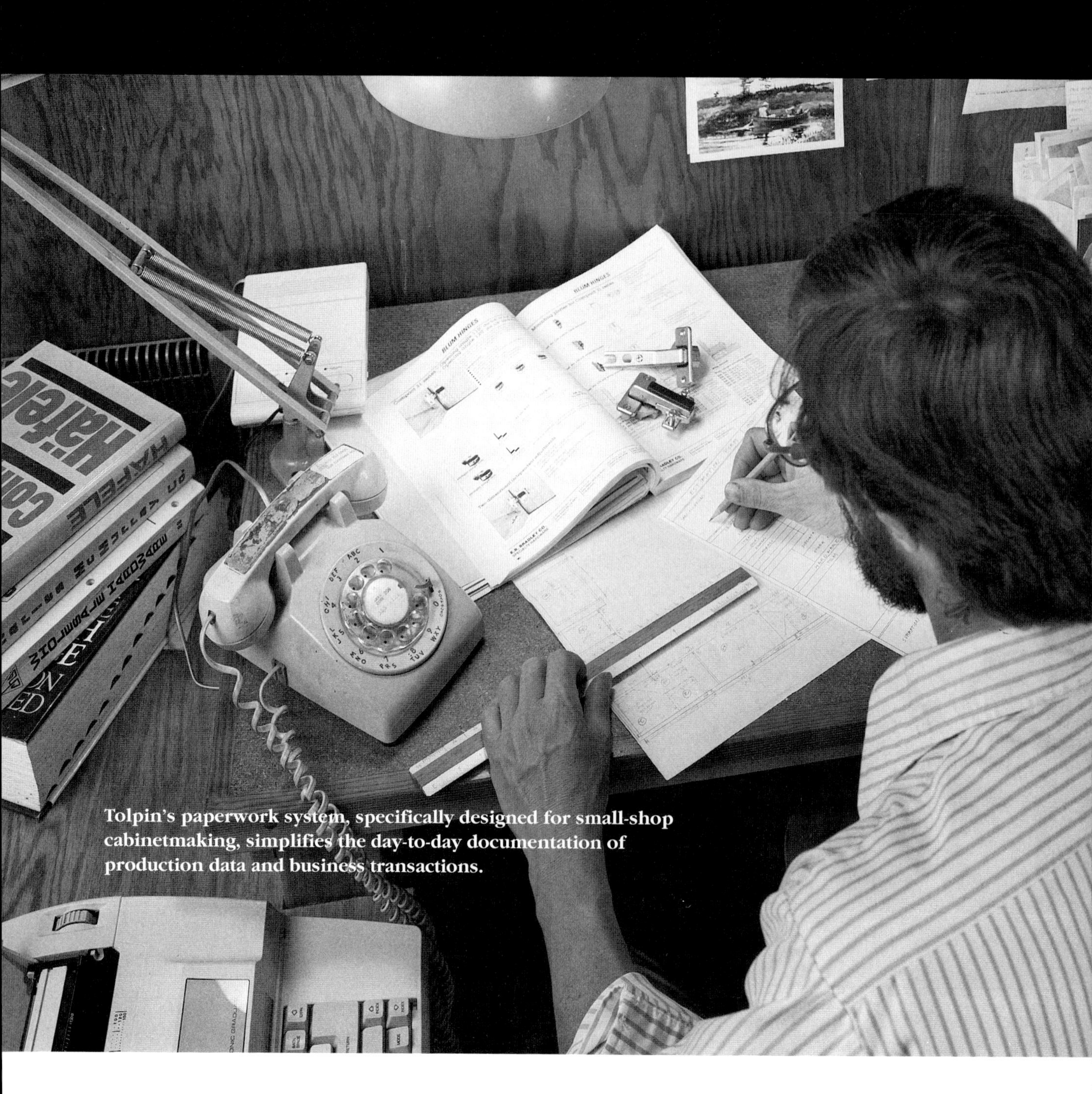

Tolpin's paperwork system, specifically designed for small-shop cabinetmaking, simplifies the day-to-day documentation of production data and business transactions.

The Business

You will probably find that setting up the workshop is one of the more enjoyable aspects of working at woodworking. Unfortunately, this activity does not a business make. Having created a well-oiled cabinetmaking machine, it is also necessary to give it something to do—you have to establish a business and find work. For many cabinetmakers, accustomed as they are to shaping things with their hands, this may be the biggest challenge of all.

At the outset, it will be necessary to examine just what you expect to get out of cabinetmaking and where you wish to go with it. There is a big difference between the avocation of woodworking and the occupation of woodworking. When you ask people for money for your work, it becomes necessary to define the shop as a legal entity. You must procure the necessary permits and licenses, and register with the Internal Revenue Service. Financial records must be carefully maintained, not only to satisfy the bureaucrats, but also for your own benefit. Written contracts should be sound for the sake of all parties involved.

The goal then becomes to get the business to generate work for itself. You must provide a product or service for a decent-sized pool of consumers at a price they are willing to pay. Thus one of your most necessary, most difficult jobs will be to define your market—who they are, where they are, what they like to buy and how much they can afford. In addition, you will have to learn how to draw a modest but steady portion of that market to your product.

Once you've carved out your niche, it becomes critically important to foster the growth of a good reputation. Contrary to advertising propaganda, this precious commodity cannot be created by words alone. Only by working harmoniously with clients and trade professionals and by honoring all guarantees of craftsmanship (both written and implied) will your name become synonymous with honesty and quality. And with a name like that, the world will indeed beat a path to your door.

For many woodworkers, this might all seem too tall an order, but be assured that you already have the raw materials necessary to become a successful businessperson. In creating the cabinetmaking machine, you have engaged in deep forethought and planning, paid painstaking attention to minute details and expressed that deeply rooted desire to produce the very best results with the least amount of unnecessary effort. If the one-third of you who will be the businessperson continues to make use of these attributes, the two-thirds of you anxiously waiting in the shop will be kept busy indefinitely.

Chapter 15: Structuring the Business

The three types of legal business entity a cabinetmaker can establish are a sole proprietorship, a partnership or a corporation. Small businesses usually start out with a sole proprietorship because it's the easiest to set up and maintain—the cabinetmaker works alone, receives all the profits (and incurs all the liabilities) and has the right to sell out or terminate the business at will.

Entering into a partnership or a corporation is usually considered in response to a perceived need to grow. Although the temptation to do so is very great from the standpoint of profit, growth usually entails certain sacrifices. The mere act of taking on employees, for example, can quickly transform a formerly independent woodworker into a woodmonger and an administrator, whose primary responsibility is the management of people rather than projects. There are other complications. The employee will have to increase shop income to the point where you can afford to pay a decent wage plus an additional 30% for workmen's compensation insurance, federal unemployment tax, your share of the social-security tax and an increased liability-insurance premium. Remember that paying these bills, filing quarterly statements, preparing payroll ledgers and managing the employee will be done on your own time. Finally, if you hire one or more employees, you must run your shop in accordance with the equipment and safety standards formulated by the Occupational Safety and Health Act (OSHA), or risk a hefty fine.

The second way to structure the business—the formation of a general partnership—brings another worker onto the shop floor without the complications and expense of becoming an employer. Working with another person on an equal basis is often a real shot in the arm, bringing in a burst of creative energy, new ideas and perhaps a larger clientele. But before you say "howdy partner," consider some of the disadvantages. Working productively with another person means learning to get along together, which is not always an easy task. Decisions concerning design and production routines become negotiable, and the products of the shop will no longer represent your talents alone. Indeed, in a general partnership, you are legally liable for all of the work and services of the shop. If your partner produces a lemon and a client responds with sour grapes, both your partner and you eat them.

Although there is no legal way to limit the sharing of liability, you should draw up a written agreement that will at least clearly define the relationship between the partners. In addition to defining the responsibilities of each partner, the agreement should stipulate how income will be drawn from the business, how profits will be shared, the amount of capital invested by each partner and the way in which the business may be sold (including buy-sell provisions within the partnership). Neglecting to address these issues in advance can lead to a drawn-out, often ill-natured round of negotiations later, which may be impossible to resolve without fattening the wallets of several lawyers.

Because of the increased liability, partnerships are often incorporated. (Sole proprietorships also may incorporate to protect personal assets from business-related liabilities.) As a corporation, an individual's personal liability is limited to the amount paid for his or her share of the stock. Corporations also offer distinct tax advantages, for example, the ability to deduct in full the cost of fringe benefits such as life and health insurance. (These deductions are not available to unincorporated businesses.) A type of corporation called a subchapter-S corporation is specifically tailored to small businesses: Because shareholders can claim profits (dividends) as ordinary income, double taxation is avoided. But a corporation is expensive to set up and maintain, and the reams of paperwork require professional expertise.

In the creation of any business, it's a good idea to seek the advice of qualified professionals. There are lawyers who specialize in business law, but you can also work with a certified public accountant. CPAs are usually less expensive than lawyers and often more receptive to small-scale business matters. Find a good one by asking for recommendations from other small-shop owners. An accountant will also be necessary to help you set up the books (see p. 108).

License, Registration and Reporting Requirements

Most counties and municipalities require that businesses operating within their borders be licensed. If you operate under a name other than your own, you will probably have to file for a "doing-business-as" (dba) or "fictitious-name" designation. Obtain information on these requirements from city hall or the clerk at the county courthouse.

Your state will very likely require registration of your business, especially if the government collects sales tax on the exchange or sale of tangible property (like cabinets). The state will issue a permit to sell taxable property, which also allows you to purchase raw materials and supplies tax-free. (Most wholesalers in a sales-tax state require customers to file their permit numbers with them.) In addition, your state may require you to register as an employer if you qualify as such. In either case, the state will require you to file quarterly income reports. Obtain state registration information from the taxation agencies located in the capital, or from branch offices located in the larger municipalities. Check the government section of the phone book for their addresses.

Last, but by no means least, is the federal government, which will insist on knowing all about your little business. If your company is structured as a simple sole proprietorship without employees, your social-security number serves as identification; otherwise, you are required to file for a Federal Employer Identification Number. If you have no employees, your contact with the Feds will be much simplified. Any business profit is treated simply as income on your personal income-tax return (summarized on Schedule C: "Profit or Loss From Business"). Social-Security tax liability is collected as a self-employment tax, computed on Schedule SE and filed with your 1040 return. The federal government will require quarterly payments against your tax liability if it exceeds a certain level.

If you have employees, you will be in touch with the government on a regular basis, depositing federal unemployment tax, social-security tax and withholding tax, and submitting quarterly statements of earnings.

As a corporation, your business must renew its license in the state of registration on a yearly basis and file its own tax statements independently of your personal income reports. (This is that extra paperwork I warned you about.)

Getting Help without Hiring

If you find yourself sharing the work with another cabinetmaker in your shop, legally you have to hire him or her outright or form a general partnership. But there are ways to get help in the shop without becoming an employer. There is nothing illegal about someone working there without pay (excluding minors, of course). This someone will be either crazy or your spouse (or, more than likely, both). With a spouse helping out, the business will generate more income without the headache of additional tax liability and paperwork. There may, however, be some tax advantages in hiring a spouse, although these are quickly disappearing. Consult an accountant before you decide.

Another dodge, which I occasionally use when faced with impending deadlines, is the hiring of subcontractors. The federal government defines subcontractors as people in business for themselves who sell their services to you. A subcontractor can supply you with components, such as drawers and doors, or perform special milling operations, such as running molding or sizing panels, and simply bill you for the work. (If your payments to any one subcontractor exceed a certain amount, at this writing $600, you will have to file a 1099 on the subcontractor at tax time.) As long as subcontractors work on their own time, in their own way (not subject to your management) and on their own equipment, the federal government will not recognize them as employees. If you have any questions about the specifics of this relationship, ask the Internal Revenue Service before they ask you.

Setting Up and Keeping the Books

You cannot successfully run a business of any kind, size or structure without accurately keeping track of what is coming into it and what is going out. In fact, there is no choice—the federal government requires that books be kept for any operating business. Once properly set up, however, the books are not that difficult or time-consuming (perhaps six hours a month) to keep. And well-maintained books are a good barometer of just how well business is doing on a monthly basis.

It's a good idea to have professional help when setting up the record-keeping system, and you might even want help in maintaining it. An experienced bookkeeper will do as good a job as a CPA and be less expensive. Get recommendations from owners of small businesses in your area: The more an accountant knows about small business, the more he or she can do for you.

Even with the help of a good accountant, the cabinetmaker will still be involved in the day-to-day bookkeeping. Since few accountants will enjoy seeing their clients walk into their offices every few months with a shoebox full of receipts and check stubs, you will need a good way to keep track of daily business. Standard procedure is to post income on an income ledger and outgo on an expenditure ledger. Each month the entries are summarized, so it becomes a simple matter at year's end of entering the sum of the monthly subtotals on your tax return. Although standardized ledger forms are available at any office-supply store, my versions on pgs. 109 and 110 are tailored to the independent cabinetmaking shop. When using these ledgers, be sure to post entries on a regular basis, or receipts will pile up and confusion ensue.

The system of bookkeeping I use is a single-entry system using the cash-accounting method. "Single entry" means that income and expenditures are recorded without consideration of other assets or liabilities. A cash-accounting method simply records income when it is received and expenses when they are paid; no provision is made to account for unpaid expenditures and sales credits. Although these simple accounting practices limit the extent to which the

INCOME LEDGER		Month of:		
		AMOUNT RECEIVED		
DATE	PROJECT NAME	Deposit	Final	TAX COLLECTED
1				
2				
3				
4				
5				
6				
7				
8				
9				
10				
11				
12				
13				
14				
15				
16				
17				
18				
19				
20				
21				
22				
23				
24				
25				
26				
27				
28				
29				
30				
31				
TOTAL				

overall health of the business may be assessed at any given time, they are sufficient for low-sales-volume businesses with a minimum of uncommitted inventory and a low credit load: a good description of a one-person cabinet shop.

If, however, you plan to hire personnel, prepare to deal with considerably more paperwork and the more complex accounting methods of double entry and accrual. As an employer, you will also need payroll ledgers and the complete set of federal and state tax-recording forms. See your accountant for these.

Depreciation of equipment and inventory control are two other important

EXPENDITURE LEDGER Month of:

				DISTRIBUTION																			
DATE	CHECK NO.	PAYEE	TOTAL AMOUNT	Materials and Shop Supplies	Shop Rent	Utilities (incl. Telephone)	Taxes and Licenses	Subcontractors	Advertising and Portfolio	Bank Charges	Vehicle Expenses	Dues and Publications	Freight	Insurance	Laundry	Bookkeeping and Legal Services	Office Expenses	Shop Repairs	Misc. Hand Tools and Bits	Sharpening	Work Clothes	Misc. Costs	Cash (Nondeductible)
1																							
2																							
3																							
4																							
5																							
6																							
7																							
8																							
9																							
10																							
11																							
12																							
13																							
14																							
15																							
16																							
17																							
18																							
19																							
20																							
21																							
22																							
23																							
24																							
25																							
26																							
27																							
28																							
29																							
30																							
31																							
END OF MONTH SUBTOTALS																							
Nondeductible																							

aspects of record-keeping. Work with your accountant to design the depreciation schedule and inventory-control list most suited to your specific needs.

Banking

Before a single bill is paid or a check accepted, open up a separate account specifically for business use. The only nonbusiness use of this account will be the occasional withdrawal of cash for personal use. It will be listed in accounting as a nondeductible expense.

Many banks charge business accounts considerably higher service fees than personal accounts, but there is no real reason to open a business account unless you are working under an assumed name (the "doing-business-as" designation mentioned on p. 107). Just open another personal checking account and use it strictly for the shop. If the local commercial banks won't go for that, try a savings bank or look for a credit union. These customer-owned banks usually don't charge extra for business accounts, and many don't charge for checks at all.

Immediately deposit all received income in the business account. When writing checks, always write in the checkbook's register and on the check itself the category from the expenditure ledger to which the check's recipient belongs. On a weekly basis, transfer the numbers from the register to the expenditure ledger. In this simple system of cash accounting, the income and expenditure ledgers should reconcile exactly with the checkbook balance. Any error is usually due to a slip in the recording of entries (assuming the arithmetic is correct). An hour or two at the end of each month should be sufficient to keep everything up to date, and your accountant underemployed.

Contracts

A handshake is fine when greeting potential customers, but it's not the way to seal a deal with them. If they want something from you, get it in writing: That's the only handshake that counts. Unless you have the signature of a client (signatures, if you are working for a married couple) on a document that specifies exactly what you promise to supply and for what price, you will have no recourse if the client should decide not to reimburse you for your work. You could end up being the owner of someone else's custom cabinets (not always such a bad thing—a colleague in such a situation built his new home around a nice set of cherry cabinets).

It should be a hard and fast rule in your business that no work will be performed, no materials purchased, without first receiving a signed quotation sheet from the customer. (When the quotation form is signed by all parties involved, it becomes the contract for the specified work.) In custom work a substantial deposit (at least enough to cover materials) should accompany the accepted quotation. In this way, you will have some protection even in a worst-case scenario, such as the sudden death of the client. The quotation form I've used for nearly 20 years is available through most print shops or office-supply stores; I have the print shop add my letterhead at the top, but otherwise it's unchanged (see p. 112). Buy the type of form that makes a second copy automatically, so you won't have to bother with carbon paper. In use, keep the copy and give the customer the original.

When filling out the form, accurately record the full name and billing address of the customer and the address of the site to which the products are to be delivered. Take the phone numbers for the customer's office, home and site (if different), and double-check them for accuracy. If a general contractor is involved in the project and the cabinetry is included in his or her contract, you will be submitting the quotation to the contractor. (Since the contractor adds a certain percentage to the bid, it isn't ethical for you to tell the owner the quoted

Quotation Form

Quotation

INTERWOOD

residential - commercial - marine
interior woodworking

Jim Tolpin
P.O. Box 62
Ferndale, CA 95536

SUBMITTED TO		PHONE	DATE
STREET		JOB NAME	
CITY, STATE AND ZIP CODE		JOB LOCATION	
			JOB PHONE

We hereby submit specifications and estimates for:

We Propose hereby to furnish material and labor, complete in accordance with above specifications, for the sum of:

______________________________ dollars ($______________).

Payment to be made as follows:

All material is guaranteed to be as specified. All work to be completed in a workmanlike manner according to standard practices. Any alteration or deviation from above specifications involving extra costs will be executed only upon written orders, and will become an extra charge over and above the estimate. All agreements contingent upon strikes, accidents or delays beyond our control. Owner to carry fire, tornado and other necessary insurance. Our workers are fully covered by Workmen's Compensation Insurance.

____
Authorized Signature

Note: This may be withdrawn by us if not accepted within______days.

Acceptance of Quotation - The above prices, specifications and conditions are satisfactory and are hereby accepted. You are authorized to do the work as specified. Payment will be made as outlined above.

Date of Acceptance:______________________________

Signature______________________________

Signature______________________________

Receipt Form

RECEIPT　DATE ________ 19 ___　**No.** 3100

RECEIVED FROM ____________________

ADDRESS ____________________

____________________ DOLLARS $ ________

FOR ____________________

ACCOUNT		HOW PAID	
AMT. OF ACCOUNT		CASH	
AMT. PAID		CHECK	
BALANCE DUE		MONEY ORDER	

BY ____________________

amount.) But if the cabinetry has been omitted from the general contract, so that the owner can shop and contract for it independently, you will be submitting the quote to the owner directly. If you're not certain how matters stand, ask.

The main body of the quotation form is for the description of the product, which must be precise. Include the grade and type of materials that will be used, the style of the product (referenced to a sample number), specific sizes and configurations and the type of finish. Be careful not to infer the inclusion of items or processes that are not being quoted, such as counter surfaces and their installation. If there is not enough room on the form, don't condense the descriptions; instead, use another sheet and note its existence on the first page. Have the customer initial your copy of the additional specification sheet.

Immediately below the product description is the space in which to present the bid itself. (See p. 118 for advice on pricing.) As when issuing a check, write out the amount in full. If your state has a sales tax, specify "plus tax" after the amount. I generally avoid including sales tax in a bid because it is too easy for a client to perceive my bid as high in comparison with others where sales tax is not included. The payment schedule is up to you and the client. I usually ask for a third down on acceptance of the quotation, a third halfway through the project and the final third on delivery. Avoid specifying the date of delivery in the contract unless forced to do so, and since "delivery" means different things to different people, specify whether the cabinets are f.o.b. your shop (freight on board, i.e., loaded on a truck) or freight paid (delivered to the site).

Acknowledge receipt of a client's check with a standard receipt form, like the one shown above. I use a version made by Rediform, which provides two extra copies. I put one in the client's file, which also contains my copy of the contract, and the other with my income ledger. Note that this type of receipt also provides a space to show the amount of the account and the balance due. This eliminates the need for a separate invoice form to keep track of the account. Look for the form at your local office-supply store.

Note that installation is a separate matter entirely. In some states, you need to be a licensed subcontractor to perform any work outside your shop. In others, there are specified limits. If you intend to do installations (and you probably will, to ensure the best results), check with your state's labor department or contractor's licensing board to see what is required. In any case, it is never wise to include installation within the terms of the original quotation. Installation of custom cabinetwork is notorious for unpredictability and labor overruns. If at all possible, negotiate the contract for installation on a cost-plus basis.

A cost-plus contract can take a variety of forms, so specify to the client which of them you mean. One standard version is to charge the cost of labor, plus a 10% markup on materials. Another version groups labor and materials, and adds a specified rate of profit. A customer will sometimes ask that a ceiling be placed on the contract. In this case, you're probably better off going back to a fixed bid at the top end of your personal estimate. I try very hard to avoid ceilings, as I usually find myself working on top of them—without pay of course.

There is often the need to write specific clauses into the quotation, sometimes referred to as jump-out clauses, to protect yourself from legitimate contingencies. The generic quotation form shown on p. 112 already lists a number of these clauses: the requirement of a written order to initiate any change in the quoted specifications; freedom from liability due to delays beyond your control; and the requirement of the owner to carry his own site insurance. Other contingencies, if predicted, should be covered in a separate clause. These might include the right to substitute materials if the specified materials are unavailable or would unduly delay the project; the right to change the quoted price due to the substitution of items at the client's request; and the exemption of liability if the owner supplies or requests unconventional or unknown materials (particularly finishing products). In general, if you can foresee the possibility of any contingency that will affect your control over the final product, get it down in writing. If you should think of one after the contract has been signed, pen it in on all copies and have the client initial it.

The signed quotation form (now a contract) offers a basic guarantee to the purchaser that the product you supply will be the one specified, and that it will be constructed in a workmanlike manner according to standard practices. If you wish to add to this rather nebulous promise, that is entirely up to you. A time frame within which any defects in workmanship will be corrected without charge can be stipulated in writing. But I find it better to leave this unwritten and open. It seems to me that specifying a 30-, 60- or 90-day limit implies that the work will fall apart the day after.

Finally, be sure that your quotation, once accepted, is complete. If the client accepts your bid with the specified deposit and a hearty handshake, you still do not have a contract. Look at the bottom line—the one provided for the client's signature; it is this that makes the document legal and binding, and it had better be on it. If the unthinkable should occur and the client reneges on subsequent payments, the terms of the contract ensure that the products still belong to you and can be liquidated to cover your investment in them. If, however, you slipped up and delivered the cabinets without taking the final payment, your only recourse may be to place a mechanic's lien on the client's property and take him or her to small-claims court. You can consult with an attorney, or do it yourself after reading up on it (see Appendix II on pp. 143).

Insurance

Unfortunately, there are innumerable other things beyond our control besides a client's disinclination to pay a bill. Fortunately, we have a booming insurance industry in this country ready and willing (for a price, of course) to protect us from the vagaries of nature and the inhumanity of man. To protect business and personal assets, cabinetmakers need to purchase two basic forms of business insurance: insurance against loss and liability insurance. If there are employees, workmen's compensation and disability insurance must also be obtained. Consult your state's labor board to ascertain from whom this coverage is purchased and the amounts required.

Protection against loss due to fire will be provided in a three-part package: "Basic" insures your equipment, inventory and the building if you own it; "legal liability" covers a rented building and is necessary if your landlord won't include a "hold harmless" clause in your rental agreement; and "property damage" insures other sections of the building not occupied by the business. Optional "extended coverage" protects against loss from causes other than fire, such as smoke damage, storms, explosion, vandalism and theft. The premiums for these policies, which are relatively high for woodworking businesses, can be reduced somewhat by installing dust-collection and sprinkler systems, and by using nonflammable finishing materials.

Liability insurance offers protection from lawsuits in the event that a non-employee sustains an injury while in your shop. (Have you ever watched helplessly as a curious customer picks up a gleaming 2-in. chisel, says "This looks sharp," and proceeds to slice off a chunk of thumb?) Of course, liability also protects customers if they should suffer from any ineptitude on your part (for example, when you show a client how sharp the chisel is by slicing off the end of his or her thumb). "Product liability," sometimes called "works completed coverage," covers you in the event that your product, once beyond the shop's doors, somehow manages to injure somebody or cause damage. Don't laugh: I once had a lazy Susan go berserk, literally flying off her hinges and strewing $75 worth of exotic spices across my client's kitchen floor—if someone had slipped on them, it wouldn't have been anything to sneeze at. Liability premiums depend on the amount of coverage desired, the projected gross annual income and the number of employees retained. Since I have no employees and do not operate as a subcontractor (I am considered a "vendor" in California), I have not had to buy workmen's compensation insurance or disability. In addition to a basic package of loss and liability insurance, I purchase my own medical coverage. To keep the rate on the latter down, I accept a large initial deductible.

For all these insurances, select a broker who has experience insuring woodworking shops. A knowledgeable broker will help determine the amount of coverage you need and clarify the details of the coverage as it relates to the woodworking business. Avoid brokers with no commercial experience; they will not know how to specify the details of your operation nor how to ask the right questions to obtain the lowest rates.

Chapter 16: Market Analysis, Product Design and Pricing

The business of the custom cabinetmaker is to supply people with cabinetry designed to meet specific needs. These needs can range from the purely aesthetic to the strictly pragmatic. To become a viable part of this marketplace, the cabinetmaker must learn who the clients are and how to find them, and develop an appealing product that can be built efficiently. The cabinetmaker must also price this work in a way that will ensure not only survival, but also financial success.

The Clientele

The continued survival of the custom cabinetmaking business attests to the existence of aware, appreciative consumers. However, this clientele is neither evenly distributed across the country nor homogeneous in character. It is easy to assume that the patrons of the custom shop come from the upper-income levels of the population, but I have discovered that there is considerably more to the demography of our clientele than money. Most clients I've dealt with over the past two decades share the following common traits (in descending order of frequency): an attraction to the beauty of wood and other natural materials; a strong appreciation for fine craftsmanship; a good education, often through college level; a disposable income beyond the bare essentials; the ability to verbalize clearly, if not illustrate, the product they wish to have built for them; and a desire to interact with the craftsman behind the work.

Where do you find these people? You should begin by looking in areas that offer something to them. Regions with a strong tradition of woodworking usually boast a large, appreciative group of potential consumers. The New England states; parts of New York, Pennsylvania and Kentucky; the upper Midwest; the north coast of California; and parts of Arizona and New Mexico readily come to mind. Regions with colleges, universities and high-tech industries tend to have larger pools of potential clients. Other areas of the country, because of outstanding beauty and attractions such as large ski or golf resorts, seem to attract people who are eager to dispose of their disposable income.

But for a region to be of interest to custom cabinetmakers, it must not only appeal to a promising clientele, it must also offer prospects for future growth. You won't be employed for long if there is a dearth of homes into which to put your products. (As Neil Young so aptly put it, "You can't have a cupboard if you ain't got a wall.") In general, the healthier the construction trade in a region the larger will be the trade in custom cabinetwork.

The Marketplace

To develop a clear picture of construction activity in an area, begin by looking at the building permits issued over the past several years. These documents are on public record at the county's building or planning department. You are searching for a substantial number of large single-family homes with a high cost-per-square-foot value, and for owners who are also listed as builders. These are good indicators of custom homes, which probably used custom cabinetry. Of course, a strong showing of kitchen remodels is also a promising sign. A talk with several realtors who list custom homes and a meeting with a representative of the local builders' association should help complete the picture of the community's current vitality and give you some insight into the future as well.

If construction in a region is doing well, it's a good bet that there is considerable competition for the work, unless the growth is a recent phenomenon. You should never be deterred because there are lots of cabinetmakers in the region (I'll talk more about this under marketing in the next chapter), but do try to get a handle on the intensity of the competition. Divide the number of consumers by the number of suppliers; if the building permits reveal that 300 new homes were built last year, and if you assume they all procured their cabinets from the 25 cabinet shops listed in the local Yellow Pages, each shop had the potential for supplying a dozen homes with cabinets. Unless these shops were all one-person operations, this is not a lot of work per shop. This kind of market analysis can be misleading, however, as not all custom work results from the construction of new homes and not all shops seek out custom homes. For shops geared to the high-volume production of modularized cabinetry, custom work is all too often a losing proposition. These shops make their money supplying tract homes and multiplexes.

To get a more accurate feel for the competition, go out in the field and visit the local cabinet shops oriented toward custom residential work. It won't take a Sherlock Holmes to determine how many are competing directly for your market, and how pressed they are to meet the demand.

To complete your understanding of a region's marketplace, research the types of products that are in vogue. Get a feeling for the area's prevalent trends in cabinet styling, configurations and finishes by visiting new homes on the market or viewing photographs of interiors on file at local real-estate offices. Talking with local interior designers, residential architects and custom-home builders will broaden the scope of your inquiry. It is not necessary, of course, to be an absolute conformist to succeed in this business. But if the style in which you work falls outside the current trends, you may want to rethink your product.

Product Definition

For the successful introduction of a product into a marketplace, the product must have widespread recognition and appeal. If you were to insist on marketing traditional "New England pine" cabinetry in an area like coastal California, your response would come only in the form of innumerable blank stares. In this part of the country, a cabinetmaker is much better off working in the style of traditional California craftsmen and designers such as Greene and Greene. In the Northwest, cabinetwork with a delicate, Japanese style finds widespread favor, as does the Taos style in the Southwest. In my experience, it is far less frustrating (and healthier economically) either to work within the parameters of a regional flavor or to strike out and develop an entirely unique style, than to try to sell a traditional style that has no local appeal.

Once you've developed a style that sells and that you enjoy producing, stick with it. There are two good reasons to do so: product recognition and production efficiency. It will not take many years before your shop and the work it produces become synonymous in the collective mind of the marketplace, and people will come to you for that kind of work. The second reason, production efficiency, is what allows you to earn the butter for your bread. If you have to change setups and processes every time you get a job, your shop is not running efficiently. Once your shop is set up, your time should go primarily to the production of the product.

There are two further ways to define yourself and your product to the public. The first strategy is to specialize in one type of cabinetwork, such as wall systems or kitchens. You can bill yourself as a specialist ("Mr. Kitchen" of the Hoboken Kitchen Design Center, for example) and focus production and marketing strategies to take full advantage of a narrowed field of endeavor. The second strategy is to enter the market at the top end. The competition is usually sparser for work at the leading edges of the market, but if your product doesn't truly belong there, you could end up running home with your tail between your legs. Since either of these strategies automatically reduces the pool of prospective customers, employ them only in a healthy marketplace with strong growth potential.

Pricing the Product

Once you have targeted the clientele and developed an appealing, well-built product, you must then figure out how to price it fairly. Both overcharging the consumer and shortchanging yourself are ultimately fatal errors. To price accurately, you must clearly understand the factors that determine a product's price, then develop an estimating system that will quickly and accurately price the work in a way that maximizes its appeal to the client yet leaves you a reasonable margin of profit.

The price of a product is determined by the cost of the materials and shop supplies consumed, plus the expense of shop time and outside subcontracts. This sum is then multiplied by a factor to produce a profit margin beyond cost. Sound simple? It actually can be quite complex. Let's take a close look at the components of this formula.

The materials cost is based on the total amount of wood products consumed by the project, with a 15% markup to account for production waste. Add to this the cost of fasteners, finishing materials and hardware. Shop supplies include materials consumed in portions (glue, filler putties, solvents and thinners) and disposable items (rags, sandpaper and throwaway brushes).

Determine the hourly shop rate by adding the shop overhead (computed on a per-hour basis) to an hourly labor rate. To figure the shop's overhead, refer to the expenditure ledger (see p. 110), which accounts for most monthly expenses. If you are just starting up, base your expenses on projections. Be sure to account for the following items: rent or mortgage; repair and maintenance; utilities, including telephone; vehicle expenses; machinery maintenance; business expenses such as insurance, office supplies and license renewal fees; advertising and portfolio development; bookkeeping and tax-preparation fees; and such miscellaneous expenses as work clothing, laundry and hand-tool replacement. To be really accurate, go beyond the expenditure ledger and account for equipment depreciation and losses such as bad debts and projected inflation on basic shop expenses (including major tool replacement).

The final per-hour overhead rate is determined by summing all of the expenses incurred by the shop over a certain number of hours, and then dividing by that number of hours. Using the example shown on p. 119, if in one month of operation (160 hours) expenses totaled $622, plus another $58 in depreciation and other losses, the shop overhead rate would be figured by dividing $680 by 160 hours, resulting in a rate of $4.25 per hour.

That was easy; the hard part is determining the amount of money you need to make an hour. If you are the head of a household and the shop is the sole source of income, you will be asking for as high a labor rate as possible. If you are independently wealthy and custom cabinetmaking is just a hobby, you can reflect that in low product costs or by building on speculation. For the sake of us householders, please refrain from the former and have fun with the latter—a woodworker's dream is to have time to build pieces on speculation.

The point is that the labor rate can range from zero to any number that will provide a decent living for you and your family. Of course, you must give to the business in relation to what you draw out of it. To earn a high labor rate without pricing your products out of the market, your shop must not only be a well-oiled cabinetmaking machine, but you must be one sparking spark plug, both in the shop and in the marketplace.

For example, assume you decide to draw $630 per week (before taxes) from the business for personal expenses, or $15.75 per hour for a 40-hour week. Adding this $15.75 to the previously determined $4.25 overhead rate yields a shop rate of $20 per hour of operation. With this figure in hand, it's a simple matter to price a particular product. If, for example, the materials for a hutch, including shop supplies, amounted to $500, a subcontractor charged $150 for a set of doors and the hutch required 50 hours of shop time, the pricing formula would look like this: $500 + $150 + (50 x $20), resulting in a product cost of $1,650 plus tax. Add profit if you want it; if you don't, your shop does. A profit margin provides capital for reinvestment, not necessarily for growth but for the upgrading of major stationary machinery. (Only the replacement of existing tooling is captured in the shop overhead rate, which includes loss due to depreciation.) In addition, the accumulation of profit cushions the shop through periods of production downtime caused by machinery glitches or your absence from the shop floor, as you do bookwork, talk on the telephone and woo prospective clients.

Determining Shop Rate Based on Monthly Expenses

EXPENSES PER MONTH	
FIXED	
Shop rent	$250.00
Waste collection and water	14.00
Equipment depreciation	58.00
AVERAGED	
Shop maintenance	45.00
Telephone	30.00
Electricity	30.00
Business expenses (insurance, licenses, office supplies, bookkeeping, advertising)	158.00
Vehicle	95.00
Total overhead	680.00
per hour (÷160) =	4.25
Labor rate per hour	15.75
Total hourly shop rate	$20.00

The actual amount of profit, like the labor rate, is rather arbitrary. And profit is not always linked solely to pricing, as it can be generated by increased production efficiency and savings on materials. I suggest, however, that you add a bare minimum of 15% to the product cost. Double that amount is better if it allows your product to remain competitive. This extra percentage will provide a cushion if you need to adjust your prices in response to market conditions. Going back to the hutch bill of $1,650, the addition of a 15% profit will up the total cost to $1,897.50. To simplify the arithmetic, the profit margin can be expressed as a factor of multiplication, that is, as 1.15.

JOB CARD	PROJECT NAME:

SPECIFICATIONS

Dimensions:	Wood type:	Special hardware:	Other:

MATERIALS					LABOR		
Date drawn	Qty.	Description	Price @	Cost	Date	Operation	Hours
			Waste factor on wood (15%)				
			MATERIALS TOTAL			TOTAL SHOP HOURS	

SUMMATION

SUBTOTAL MATERIALS		NET COST	
SUBCONTRACTS		PROFIT (margin ____)	
SHOP HOURS @ ________		PROJECT COST	

Cost Estimating

While it is easy to figure out how much a product costs after it has been completed, it's the rare customer indeed who will wait until the end of the job to find out what the bill is going to be. Thus the custom cabinetmaker is doomed to give bids, knowing only the basic specifications of a proposed project.

The key to an accurate bid is a careful estimation of material and labor costs, based on a structured assessment of previous work. This task is much simplified if all the production data for every product built are carefully recorded on job cards designed for this purpose (see the facing page). I keep a copy of the job card on the clipboard that hangs on my tool caddy Moe (see p. 32). Fill in the top of the card with the name of the project (usually the client's last name, followed by a number if it is part of a larger project) and its specifications. (Include all external dimensions, the type of wood products used and any hardware other than the typical fastenings, hinges and drawer slides.) The balance of the job card is a ledger with two categories. The first category describes the materials consumed in the project, including an estimate of shop supplies and their costs; the 15% waste factor appears at the bottom of the column. The second category lists hours of shop time. The abbreviations listed on p. 122 identify the production processes discussed in the previous section. The bottom part of the card summarizes the subtotals and gives the cost of the project.

For a tool to be effective, it must be used properly, and a job card is no exception. Make its use habitual. Without fail, log all materials drawn from stock and the labor time of each process. I use a punch clock, designed by Stanley Rill of Hadlock, Washington, to track elapsed time up to a total of 12 hours (see the drawing on p. 123). Reset the clock to 12:00 at the beginning of each process and activate it with the switch. (The light indicates when the clock is counting.) Switch off the clock when the process is completed or any time work is interrupted.

Pricing Formulas—It will not take very long for you to appreciate the incredible usefulness of these job cards. Armed with a sheaf of cards, you will soon develop a knack for speedy, accurate cost estimates.

The formulas are derived directly from the information tabulated on the job cards and are based on the cost per square (or linear) foot of the products. In general, the standard lowers and uppers of kitchen casework, as well as other varieties of counter-height cabinetry, can be accurately priced with a linear-footage formula. Larger cabinets (such as floor-to-ceiling oven units and pantries) and freestanding units (such as china hutches and entertainment centers) are more accurately computed using a square-footage formula based on the surface area of the cabinet face.

As an example, let's develop a formula for pricing out a 4½-ft. by 7½-ft. kitchen hutch. Looking at a job card for a recently completed 4-ft. wide by 7-ft. tall china cabinet, we find that the total cost of materials was $450 (including 15% for waste) and total shop time was 100 hours (at $20 an hour). The cost of that project, without profit margin, was thus $2,450 ($450 + $2,000). Dividing this figure by the face square footage (4 ft. x 7 ft. = 28 sq. ft.) gives us a square-footage cost of $87.50.

This figure, however, includes the china cabinet's doors and drawers. To arrive at a universal factor for pricing other hutches, which will likely be of different configuration, doors and drawers must be treated separately. Referring once again to prior job cards reveals that a run of 20 drawers costs $200 in materials and $400 in shop time. Division of $600 by 20 drawers yields a per-drawer cost of $30. A similar analysis of a door run in the style of the china cabinet yields a per-door cost of $24.

Abbreviations of Production Processes

Process	Abbreviation	Description
Job preparation	C	Consultation
	D/C	Design and calculations
	M/H	Material handling
Sheet-stock sizing processes	Sht-C	Carcase panels (including backs)
	Sht-Dwr	Drawer components
	Sht-Dr	Door panels
	Sht-Shf	Shelves
Sheet-stock milling processes	Sht-Prep-C	Carcase components
	Sht-Prep-Dwr	Drawer components
	Sht-Prep-Misc	Miscellaneous sheet components
Solid-stock sizing processes	Sld-Face	Face-frame components
	Sld-Dwr	Drawer faces
	Sld-Dr	Door rails and stiles
	Sld-Pan	Solid-stock door panels and cabinet sides
	Sld-Md	Moldings
	Sld-Misc	Trim, edgings
Solid-stock milling processes	Sld-Prep-Face	Face-frame Ritter process
	Sld-Prep-Dwr	Drawer-face hardware preparation and edge treatment
	Sld-Prep-Dr	Edge treatment of rails and stiles
	Sld-Prep-Pan	Lamination of panels, final sizing, edge treatment, other milling processes
Component-assembly processes	Comp-Face	Face-frame assembly
	Comp-Dwr	Drawer assembly
	Comp-Dr	Door assembly
	Comp-Shf	Shelf edging
	Comp-C	Carcase panel edging
	Comp-Misc	Sub-unit assemblies (moldings, trim packs)
Final assemblies	Ambly-C	Case assembly
	Ambly-Face	Installation of face to case
	Ambly-Hwr	Hardware installation
	Ambly-Mld	Installation of molding and trim packs
Finishing	Fin-Prep	Sanding and scraping processes
	Fin	Application of finishes, including buffing out
Delivery	Del	Delivery
Installation	Inst	Installation of cabinet units
	Inst-Tm	Trimming out

Since the china cabinet had two drawers (worth $60) and eight doors (worth $192), we can subtract $252 from the total cost of $2,450. Dividing this new figure ($2,198) by 28 sq. ft. yields a cost of $78.50 per sq. ft. of face—the universal factor for hutches. Thus the 4½-ft. by 7½-ft. kitchen hutch under consideration can quickly be priced out, as shown in the estimation form on p. 124.

But what do you do when the only job ledgers you own are the blank ones you've just copied out of this book? When you are just starting out, it is, of course, very difficult to come up with universal factors. Your only solution is to ascertain what the other boys on the block are doing. I'm not suggesting you go out and get bids on a hypothetical project from your neighborhood custom cabinetmakers (you have to live with these guys). Instead, develop universal factors by analyzing the price sheets of a number of prebuilt cabinets distributed in your area. To approximate custom work, look at the models near the top of the line. By extrapolating figures for cabinets of various dimensions and configurations, with and without doors and drawers, you can begin to generate some cost formulas. It won't take you long to produce figures of your own. Remember to update your factors continually as your production skills increase.

Writing Bids—We've already seen how to present a quotation that specifies the contingencies under which the bid is written, and how payments are to be made. But what about the bottom line: the one with the money on it? You must give your price sometime, but you want to be sure you aren't cutting your throat when you do so. There's no worse feeling than when a prospective client looks at the bottom line and says: "Wow." You know at that instant that you've either just lost the job or your shirt.

Fortunately, there are ways to help keep the knife from your throat when writing bids. Begin by scrutinizing the blueprints for every factor that could affect the final cost of the project. Account for all doors, drawers and specialized hardware items, such as lazy Susans and slide-out baskets. Note cabinets with exposed end panels, because they cost more to build than those with unfinished ends; also record any unusual layouts that will affect construction methods and dimensions that will produce a lot of waste in 4x8 sheet stock.

While your nose is in the blueprints, also check that built-ins will fit through the necessary doors on installation day. (Second-floor bathroom vanities have been the Waterloo of many a custom cabinetmaker.) Also, since architects are not infallible, make sure that cabinetry does not cross windows or other wall openings. I caught this type of mistake once...but too late. The client wasn't amused by the suggestion of fitting a glass back on the cabinet that, if installed, would have marched across an expansive picture window.

But the worst situation I ever encountered was the case of the house that was "put on backwards." (Actually, it was the mirror image of the house intended for that foundation.) A mix-up in the plans with another home under construction in the development created a left-handed home with a right-handed set of kitchen cabinets. Fortunately, we were eventually able to find the cabinets a right-handed home.

There are other factors to consider before writing that final bid. If the project will demand an unusual amount of consultation or design, whether due to its inherent complexity or the intransigent nature of its owner, account for the extra time. Large projects require that more time be devoted to the logistics of handling raw materials. If a project will include delivery, check the distance and access to the site before bidding. As I mentioned earlier, never include installation of the cabinetry within the terms of a bid (unless you are forced to). If done on a fixed price, installations are a certain way to nick your throat, if not worse. Finally, give some thought to your profit margin. While a 30% markup is ideal, a lower margin might be in order if competition for the project is stiff, or if acquisition of this job will lead to more profitable work. In the latter case, a lower initial margin will serve as the "loss leader" that helps get your foot in the door. Going below a 15% margin, however, should be out of the question. You may get jobs this way, but they will not sustain your business. Remember, you work not only for yourself, but for your shop as well.

To conclude this chapter, two sample estimation forms of the type I use in my cabinet shop are presented on the following two pages. The form on p. 124 has been filled in with pricing details of the kitchen hutch that we used as an example to illustrate pricing formulas. The blank form on p. 125 can be duplicated for use in pricing your own projects.

Time Clock

The clock is wired through the switch. The light indicates that the clock is running.

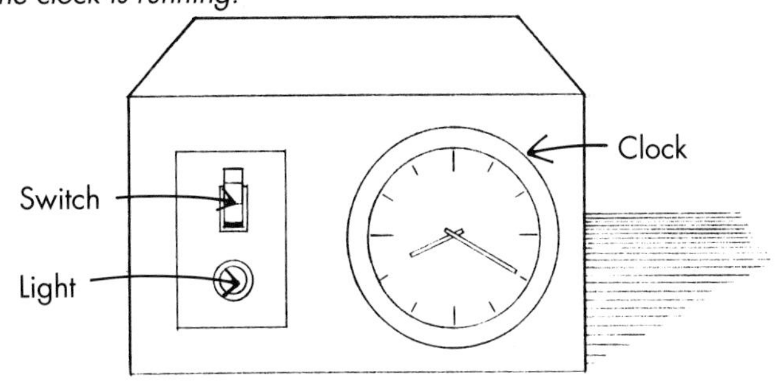

Reset the clock to 12:00 at the beginning of each time period. The hands in the sketch read 8 hrs. 20 mins. of elapsed time.

ESTIMATION FORM	PROJECT: KITCHEN HUTCH

SPECIFICATIONS:

- 4'-6" X 7'-6" X 22" (lower) 12" (upper)
- OAK THROUGHOUT
- 4 Flat recess-panel doors
- 2 Drawers
- Shelves open in upper

	FACTOR	
SQUARE FACE FOOTAGE 33.75 or LINEAL FOOTAGE	18.50	624.38

ITEM	NUMBER OR FOOTAGE	FACTOR	
Doors	4	24.00	96.00
Drawers	2	30.00	60.00
Moldings	7 LFT. CROWN	2.50	17.50
	7 LFT. COVE	1.25	8.75

SUBCONTRACTORS	N/A
OTHER COSTS (freight, delivery, etc.)	30.00

HARDWARE

ITEM	QTY.	PRICE @	COST
Drawer Slides	2	5.00	10.00
Door Hinges	4pr.	6.00	24.00
Shelf Clips	16	.15	2.40
Levelers	4	1.95	7.80
SUBTOTAL HARDWARE			44.20

Hardware total: 44.20

SUMMATION		
	TOTAL NET	880.83
	PROFIT (margin 15%)	132.12
	TOTAL (plus tax)	1012.95

ESTIMATION FORM	PROJECT:

SPECIFICATIONS:

	FACTOR	
SQUARE FACE FOOTAGE ______________ or LINEAL FOOTAGE ______________	______________	

ITEM	NUMBER OR FOOTAGE	FACTOR	
Doors			
Drawers			
Moldings			

	SUBCONTRACTORS	
	OTHER COSTS (freight, delivery, etc.)	

HARDWARE

ITEM	QTY.	PRICE @	COST
SUBTOTAL HARDWARE			

SUMMATION		
	TOTAL NET	
	PROFIT (margin ____)	
	TOTAL (plus tax)	

Chapter 17: Marketing and Sustaining the Business

When you are marketing a line of high-quality, custom-built cabinets, it is important to understand that you are selling yourself as well as your work. As we discussed in Chapter 16, the consumers of these products are not motivated by the beauty of wood and the quality of the construction alone. They are also drawn to the opportunity to interact personally with the craftsman. The expectation exists that the craftsman will reflect the same qualities inherent in his work; you are therefore expected to be as creative and exacting, and as honest and durable, as the cabinets you build. It is your task to meet these expectations in the ways in which you present yourself to your clients, not only in your business media and displays, but also in your personal etiquette.

An Overcrowded Business Card

HANDCRAFTED WOODEN CABINETRY
European Master Craftsman - Peter Eva
15 Years Experience

* Cabinet Design and Construction, All Styles
* Built-Ins and Freestanding Units
* Kitchens and Bathrooms
* Installations
* Antique and All Furniture Repair

Free Estimates
Excellent References
914-525-0822

Peter Eva
Fogginton Road
Brewster, NY 10509

Business Media and Displays

Business cards and stationery are your emissaries to the marketplace; they should not only reflect your dedication to quality but also hint at your creative talents. Graphics can be eye-catching if done well, but you should avoid cartoonish, timeworn illustrations of hand tools or furniture pieces. In design, simplest is often best, and there is no harm in simply stating your name and occupation, and the location and phone number of the shop—John Maurer's business card (shown in the display on the facing page) uses typography and an interesting layout to create an effective card. Beware, however, of overcrowding. Peter Eva's card (shown at left), while telling much about who he is and what he does, is so packed with information that it's hard to read. The descriptive material would have been better placed on the back of the card.

Sometimes it is useful to step outside of traditional business-card design and opt for an unconventional format. John Ewald's card (shown on the facing page) is oversized, measuring 3½ in. by 5¼ in. at full size; note how the broken type echoes the bamboo joints of the logo. My own card, designed with the help of artist Diane Mayers, was intended to convey to the public that I design and build traditionally inspired cabinets, furniture and entryways. People tend to remember this card as "the one that opens like a door," and the response to it has been quite favorable.

Some Effective Business Cards

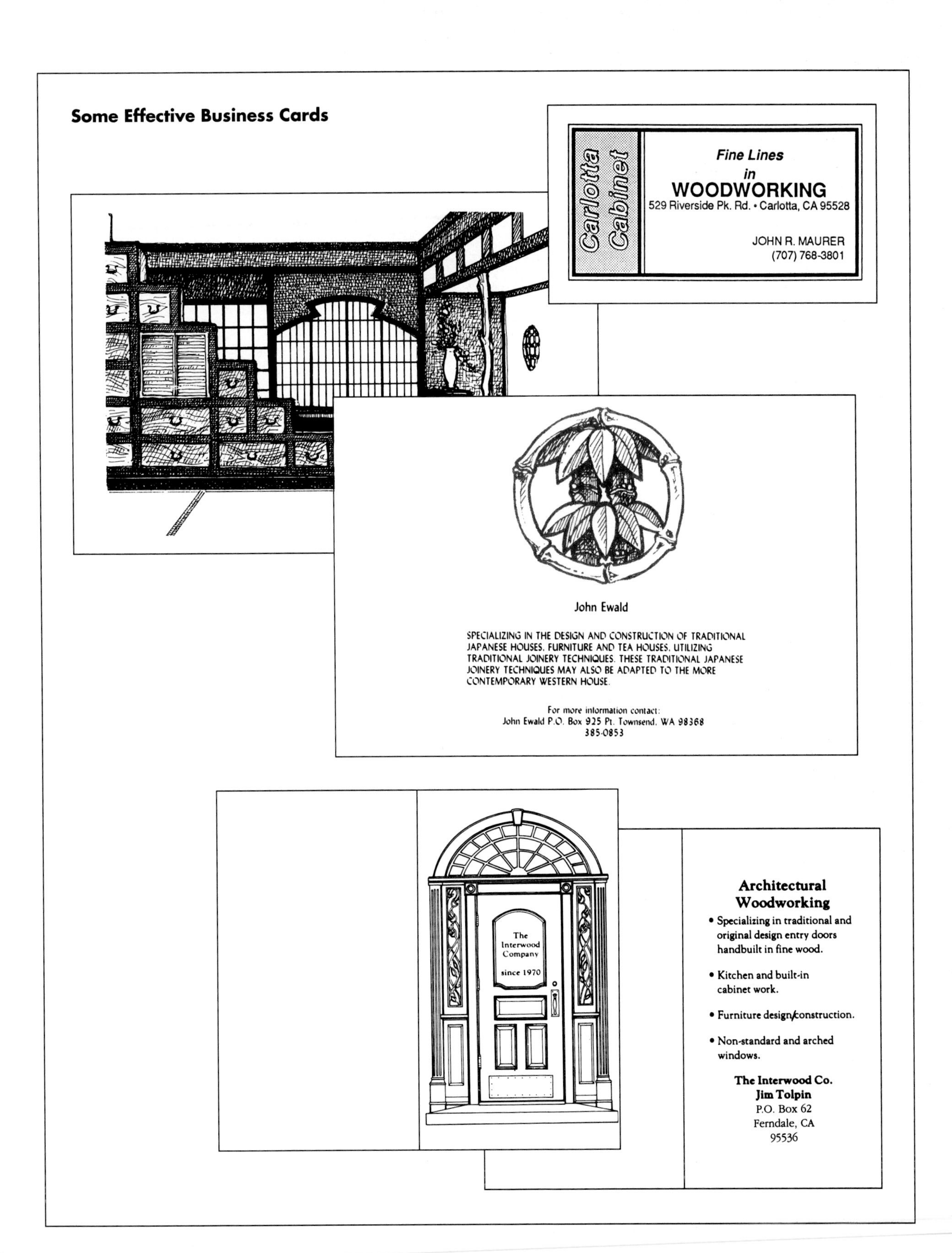

Tolpin's sample case is actually a miniature standard cabinet module. Each face displays a different door and drawer-face style and examples of pull hardware. One face is operational, opening to reveal a miniature drawer, adjustable shelf and slide-out unit. Don't get carried away with extraordinary details on your sample case; it must honestly represent your production cabinetwork.

Use the same logo and typeface for the letterhead of your stationery. Although not an essential, personalized business stationery, when used to communicate with clients and trade professionals, helps foster the impression of a meticulous, committed craftsman who means business.

Do not use or distribute this business media indiscriminately: Placing business cards at local restaurants and bowling alleys is a complete waste of time, and using company stationery for personal communication is a waste of a rather expensive material (and smacks of showmanship). Your business media should be used primarily to identify you to present and potential clients, and to communicate in a credible manner within the trade.

The rest of this chapter deals with ways of getting the market to open its doors to you; once it does, it's important that you have something besides your good looks and a business card to show off. I therefore recommend preparing a portfolio of photographs and a sample case, which represents your work in three dimensions.

My present portfolio is composed of about 20 8x10 color photographs that present an overall view of installations, and an equal number of 5x7s that highlight the details. Slides, when projected, are even more impressive than 8x10s, but the hassle and time involved in presenting them are real obstacles. (In the early days of my career, when the pickings were slim, the criteria I used to select portfolio candidates were somewhat less exclusionary than now: As long as I wasn't embarrassed by it, any piece could be included. Although the the portfolio was rather small, even then the pictures were very good.) Because prospective clients will be judging the quality of your work by the quality of your pictures, it's well worth investing in professional photography for your portfolio. Freestanding pieces are especially responsive to studio treatment.

The best photography in the world won't survive shoddy presentation, however, so be sure to buy a quality portfolio case from the local office-supply store. Ideally, the case should contain each photograph in a glare-free sleeve. Each sleeve should be attached to the binding loosely so that it will be easy to add and reshuffle photos. I like to caption each photograph with a title, such as: "East Wall, Hargrave-Garrison Kitchen." Going to these lengths reinforces your image as a creative, meticulous and detail-oriented craftsman.

A nicely constructed sample case, which is just a miniature cabinet on a swiveling base (see the photo on the facing page), will accent the "highly skilled" facet of your public image. This cabinet can feature a number of different door and drawer styles; provide examples of face-frame and non-face-frame (European) cabinet types; and demonstrate drawer slides, door hardware, adjustable shelf clips and legs (with clip-on kickboards). The sample case can also provide storage for wood samples and a selection of miniature doors that represent other available styles. Do your best work in the creation of this case, but use standard materials and procedures in its execution—it is, after all, a representation of the product you are offering to the marketplace. Don't handcut dovetails in rosewood drawer sides unless you intend to offer such features in your production units.

Building Relationships within the Trade

A custom cabinetmaker's work often comes through referrals from architects, interior designers and custom builders. For this reason, it's worthwhile to cultivate relationships with as many of these professionals as possible. When developing a list of potential contacts, it is usually most efficient to begin with architects, as they can offer referrals to other members of the trade. Since the work of many architects carries them far afield from the world of custom woodworking, you needn't call on every one of them. Eliminate any listed under "industrial designers" and "project management" in the Yellow Pages. Screen the remaining architects with a simple telephone introduction: "Hello, I am the owner of Woodchuck Woodworking. We specialize in high-quality custom cabinetwork and wonder if you have anything on the boards that might require this kind of product." Most architects will respond with a straightforward yes or no. The ones who respond immediately with a firm yes are usually involved in a residential or commercial de-

A Contact Card

CONTACT:	REFERRED BY: Architect Interior Designer Contractor Owner (Circle one)
ADDRESS PHONE (home) ________ (office) ________ (site) ________	TRANSACTION ABBREVIATIONS NA - No Answer MA - Made Appointment LM - Left Message MR - Made Referral CB - Instructed to call back UD - Update

TRANSACTIONS															
	1	2	3	4	5	6	7	8	9	10	11	12	13	14	15
DATE															

REFERRALS:

sign where they have some control over the selection of subcontractors. These are your hottest bets—respond with a well-rehearsed, brief soliloquy about yourself, and then request an opportunity to show your portfolio and sample case. A reserved yes usually means the architect is designing for builders who handle their own woodworking or subcontract it. All is not lost, however, as architects are usually willing to name the local general contractors who specialize in custom home construction or who use custom woodworking elsewhere in their businesses. They have nothing to lose by offering the names of their clients, and you have everything to gain by spreading your growing network of contacts to these professional builders.

Begin screening this select group of contractors with a telephone inquiry. Mention the name of the referring architect, then ask whether there are any projects currently under construction or in proposal that will have bids let out for custom cabinetwork. If you get a positive answer, go forth with your introductory soliloquy, and request an opportunity to show your sample case and obtain a copy of the blueprints for the project. Many contractors will say that they already have shops they feel comfortable dealing with, but will be open to meeting a new guy on the block. Take advantage of this; you can be pretty easy to work with, too.

Some architects may offer referrals to interior designers. Don't hesitate to introduce your products to them as well. You'll find, however, that most designers have little to do with interior woodworking and custom-cabinet design. They are primarily contracted to choose and coordinate wall and floor coverings, fabrics, furniture styles and other decorative aspects of interiors. But there will occasionally be a designer involved in built-in furniture designs and kitchen layouts who'll be happy to learn of craftsmen who can provide well-built products.

Contact Cards

It shouldn't be long before you are inundated with the names, addresses and phone numbers of contacts and contact referrals. You'll need to follow up all of these and develop a system by which to organize the information and keep it up to date. On the card shown on the facing page, contacts are filed into four separate categories: architects, interior designers, general contractors and owner-builders; there is also space to write in the source of the referral. (It is always a toe in the door to mention the name of a contact's colleague when introducing yourself.) A series of columns is provided in which to record the sequence of any transactions; be sure to date each transaction, and indicate its nature using the abbreviations provided. Never leave a dated transaction column blank. If an appointment was made, write it into a separate appointment calendar immediately. If referrals are offered, note them on the card and fill out a new contact card for each name.

Appointment Etiquette and a Pep Talk

When scheduling a business meeting outside the shop, avoid Mondays, which are usually hectic. Everybody is bleary-eyed from the weekend and wondering why they are at work; don't be around to make them wonder why you are there, too—they just may not care. Fridays are also bad. Everyone is anxious to go home, and you don't want to be one more reason why they can't.

Having arranged a potentially lucrative meeting, make the most of it. Be sure to arrive on time. If you're unable to keep the appointment or are running late, call as soon as possible and set up another time. This courtesy is noted and appreciated, and provides an opportunity to make a good impression before you even show up. Dress in a clean set of clothes that you keep for these occasions. Don't overdo it: You are a woodworker first, and a businessperson second. Avoid three-piece suits—go with a casual sportcoat.

With an impressive portfolio in one hand and a masterpiece of a sample case in the other, you will have some heavy artillery behind you at any business meeting. But there are two self-defeating, confidence-shaking thoughts that aspiring custom woodworkers might indulge in, which could damage that valuable first impression. The first is this: "My work stinks compared to the work of a lot of other craftsmen around here." The best way to respond to this thought is to acknowledge its truth. Nearly every woodworker at some time meets somebody whose work is better; those who feel they are the best have usually ceased to appreciate and learn from the work of others, and it shows. The other damaging thought is this: "With so many other woodworkers out there, how can there possibly be room for me?" Of course you won't know unless you try, but chances are there's plenty of room at the inn. For although there may be scores of woodworkers in the area, you are the only one currently meeting with this particular client. The opportunity for you to be in the right place at the right time puts you light years ahead of those craftsmen who are hiding in their shops (including the ones whose work is somehow always better than yours).

Once in the meeting, be as polite as you can remember how to be. Introduce yourself via an orchestrated journey through your portfolio. The photographs will provide a stimulating visual backdrop as you intone your resumé. Make your sojourn as entertaining as possible,

A Direct-Marketing Letter

Dear

I recently learned through the local builders' grapevine that you are contemplating, or have already begun, the construction of a new home or remodel. If you are interested in obtaining competitively priced, well-built cabinets for your project, you might wish to consider the services of a local custom cabinetmaker.

Employing the skills of a local craftsman allows you to obtain products that will be carefully designed and constructed to meet the specific, demanding requirements of your unique project. Nearly any configuration of cabinet can be constructed, and any of the wide variety of cabinet woods and specialty hardware items can be incorporated into the design. The choice is up to you.

I would be pleased to show you a portfolio of my work, which encompasses twenty years of supplying fine homes with fine cabinetwork in styles ranging from the strictly period reproduction to the uniquely contemporary. Please feel free to call and arrange for an appointment. I look forward to meeting you, and I wish you well on your project. Thank you for taking the time to read this letter.

Sincerely,

offering quips and interesting (but brief) sidelights to the material you are presenting. When you arrive at the end of the portfolio, turn the spotlight on the sample case. Let your host open and close the doors and drawers of the case; in so doing, he or she will discover the door and wood samples stored within. Questions and comments will invariably arise, and the prospective client will thus unwittingly take over the show. Very likely the subject of his or her own work will come up, and you should encourage this. When the time looks right, steer the conversation toward current projects on the boards and under proposal. This gives you a chance to evaluate the job potential of this particular contact.

Conclude the meeting by affirming whatever follow-up seems appropriate. If the potential for work is imminent, suggest another meeting, perhaps with your host's clients present. Even if the potential rewards of this meeting seem far off, say you will stay in touch and keep the promise—you do not want to join the crowds of shop-bound woodworkers, forever lost to the contacts you have just begun to nurture. Before exiting, hand over several of your business cards and extend thanks for the meeting. Don't dawdle, but sweep yourself out the door before the secretary tries to sweep you under the rug. As soon as possible, write down everything pertinent to the meeting. Log the necessary follow-up on the contact card; note if any other follow-up is necessary (such as sending 5x7 copies of photos in your portfolio); fill out new contact cards for any referrals; and enter subsequent appointments onto your calendar.

Direct Marketing to Consumers

It is likely that a large portion of your work will come through direct contact with the cabinet-buying public. Through relationships with the trade professionals, you will occasionally receive direct referrals to their clients. These are always good prospects, as the names of these individuals would not have been released unless they had reserved the right to contract for their own cabinetry.

Another way to learn of these self-contracting owner-builders is to comb through recently issued building permits at the county's planning department. If the project is to be built by the owner, the permit will say so. A more convenient strategy may be to join the local builders' association; up-to-date listings of building permits are often included in the newsletter. There are, of course, other benefits to becoming a member of these organizations, not least of which is

A Postcard Mailer

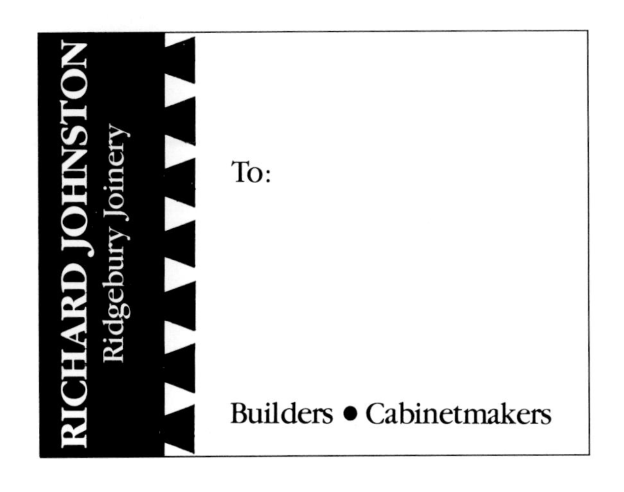

Let us create a unique environment in your home through efficient use of attics, basements and other unused or poorly designed areas

Complete Woodworking Services for the Architect and Homeowner

- Additions - Renovations - Restorations
- Custom Interior Woodworking
- Fine Furniture-Cabinetry
- Kitchens - Baths - Built-Ins
- Timber Framing

549 Ridgebury Road, Ridgefield, Ct. 06877
Phone (203) 748-0446

the dissemination of your name among a receptive group of tradespeople.

About once a month, peruse the permit data and send a business card and standard letter of introduction (see the sample letter on the facing page) to likely candidates. The postcard mailer shown above is another excellent option. Select the lucky recipients of your communiqué based on the category of the permit, the value of the project per square foot and the location. In my own estimation, if the category is residential, if the value of the project reflects an above-average investment and if the location is in an area of high land values or other custom homes, then I've got a hot prospect. After sending the letter or postcard, wait about a week and follow up with a phone call. Politely inquire whether they received your mailing and whether they might wish to meet with you to discuss the project further. Don't be surprised if they say no—the rate of positive return on any kind of marketing strategy is, almost without exception, far less than 10%. They will tell you they have already hired a cabinetmaker, or they are doing the cabinets themselves with the jigsaw they got for Christmas (don't laugh) or they are planning to order them from Sears. Good for them. For the one in twenty who says, "Yes, I'm glad you called," pull out a fresh contact card and fill in the name, phone number and address (and the site address if different), and indicate in the transaction column what the next move is to be. If a meeting is planned, note it on the card and on your appointment calendar.

The decision whether to meet clients at their home or in your shop is generally left up to you. Most people, however, will express a desire to see where you work. As discussed earlier, the public seems to carry a universal image of the venerable woodworking craftsman, and they are curious to see one in action. Don't disappoint them—invite them to your shop.

Schedule the meeting to take place during a logical break in your workday, such as at lunchtime or immediately after quitting time. Try to have the shop somewhat neat so your prospects won't trip over wood scraps, but leave some hand tools out to satisfy their expectations of Geppetto's workshop. Once they arrive, give them the shop tour and briefly mention the work in progress. As in your meetings with the trade professionals, your top guns are your portfolio and sample case. Steer the clients to the office (if you have one) and make them comfortable. Offer tea or coffee, and then jump right into the show.

You can determine quite a bit about your prospects by the questions that arise. If they all relate to money, a yellow light should switch on in your head; it's obvious the clients are less interested in the product than in what you charge for it. But this doesn't necessarily mean all bets are off. It can simply indicate they are pricing out the various components of their new home. In any case, spend less time on product information and get right down to the bottom line.

It doesn't hurt to bring up the subject of money. In fact, I usually tell people the approximate costs of the projects shown in the portfolio. I can then judge from their reactions just how serious they are about investing money in their dream cabinetwork. You certainly don't want to be the one who shatters these dreams, but you don't want to be part of a nightmare, either.

A type of prospective client I am usually wary of is the one who concentrates solely on the work, with absolutely no interest in cost. Although it might appear that the ideal "cost-is-no-object" consumer has come to your door, in reality this person is probably there to pick your brain and get some free ideas. But never write off any client with whom you meet; simply get down to brass tacks and arrange to develop a bid for the project. If the client cannot leave a set of plans, arrange for copies to be made at a shop where you have an account. The copies can be left for you to pick up at your convenience. (Clients usually pay for the copies, but if it's charged to your account, it will likely still be cheaper than having to mail the original prints back to the owner.)

Be prompt with any quotation and send it along with a note on another business card thanking the prospective client for coming to your shop. Don't hold your breath awaiting a response; it may take months, and it may not come at all. It won't accomplish anything to follow up the quotation with a phone call, either. If the job is yours, which is all you really want to know, you will be told in due time.

Advertising Avenues

Many people in the trades will tell you that trying to find work through advertising is a dead-end street. In my experience, they are almost right, especially about paid advertising. But even people who live on dead-end streets put things inside cabinets. If you are just getting started or have recently moved your shop, it is worth trying to reach them through advertising.

The least expensive way to buy advertising is to place a classified ad in the local newspaper. You can spend a little more money for a display ad, but I don't believe these are any more effective. Newspapers usually offer a break to long-running advertisers (which is the only effective way to use this kind of advertising) and often provide a separate column for service listings, including cabinetmaking. See if you can get them to add "custom" to the title. For a while, you may be the only one listed under this heading.

The ad I usually run is shown on this page. It contains only 16 words plus a phone number, yet conveys enough information to introduce the products I make. Placing my name and occupation in bold print at the beginning of the ad leaves nothing to chance. Over an extended period of time, the classified readership subconsciously links your name to cabinetmaking, and this will occasionally produce an inquiry. But don't pin your hopes on it.

Another way to buy advertising at a reasonable rate is to list your business in the Yellow Pages of the region's phone books. For those people who are new to an area or who have no source of referrals, the Yellow Pages serve as a credible introduction to the service community. If your name is there, you may get a call. Always ask callers where they got your name—it can be reassuring to know that your advertising dollars have some effect.

Probably the most effective advertising is the kind you don't have to pay for. I call it indirect advertising, and define it as anything that ties your name to your work. Most communities offer a host of opportunities for this type of advertising. You can participate officially in community affairs that enjoy high public visibility, such as parades and fairs; you can donate shop time to nonprofit community causes; and, if you have the inclination, you can teach woodworking to kids on a volunteer basis, or to adults at night school for pay. For a number of years, I opened my shop several evenings a week to a half-dozen amateur woodworkers. I charged by the hour for the use of the shop, throwing in advice and occasional instruction at no extra charge. Numerous referrals and subsequent jobs resulted from these contacts. I've also helped build playground equipment, floats for parades and small boats for raffles. All these activities can help make your name synonymous with your work and can set you up for a feature story in the weekend edition of the local newspaper.

Also consider joining the local Chamber of Commerce and other business organizations that promote local industry. Not only will this enhance your credibility as a business entity within the community, but participation in these officially sanctioned groups will also spread word of who you are and what you do to a large number of professionally and socially active people.

A Classified Ad

Jim Tolpin, Cabinetmaker:
Finely crafted custom kitchens, baths, furniture and entryways for fine homes since 1970. 768-3975.

A Good Reputation

The single most important asset that a person can acquire in this business, other than a mastery of vocational skills, is a good reputation. When you are perceived by the community as having this near-mystical quality, your security as a self-employed craftsman is greatly enhanced. This esteemed stature becomes the only form of advertising necessary to attract a steady flow of business to your shop. There are many ways to cultivate a good reputation within your community. The secret lies in learning how to conduct yourself with both the people you work for and those you work with.

I must admit that I have at times paid dearly to keep the "good" in my "good reputation." Indeed, I have found a great deal of truth in that old maxim that proclaims "the customer is always right," even on those occasions when I knew I wasn't wrong.

With almost every client there are discrepancies that somehow slip between the cracks of mutual understanding—times when a drawer will not be where it's supposed to be, or when proportions of moldings seen full scale do not match the client's expectations. I find that the best way to handle situations like these is simply to correct the "error," if it can be done without major expenditure of additional labor. This allows the clients to feel not so much that they are right, but that you are truly concerned with their happiness.

I suggest that on any job your goal should be to leave the clients feeling good about you and your work. It helps to remember that it is not just this client you are pleasing, but all those clients in your future. In the end, it is best to have former clients speak highly of you to their peers. It is beneficial if they speak of you at all, but absolutely deadly if they speak ill of you. The effects of ill will at the end of a job are long-lasting and far-reaching, and will be far more detrimental than swallowing a little ego and eating humble pie. If you have been careful with the written agreements, you should never have to eat too heartily.

You'll be relieved to hear, however, that there are less stressful ways to help make a client feel good. You will, of course, always be courteous when customers walk into your shop, but you must also be attentive to why they are there. Although there is a limit to the amount of time that can be taken away from production without charging for it, your clients need to feel at ease talking with you. My own rule of thumb is that when the clock moves past 15 minutes, it's time to schedule a paid consultation. If an unannounced, unknown visitor comes to your shop, graciously suggest scheduling a meeting at a mutually convenient time. If the inquiry begins to resemble a consultation, letting it be known that your consultation fees are so much an hour will separate the wheat from the chaff.

As subsequent meetings with clients take place, always show interest and respect for their ideas on design, layout and, heaven forbid, construction techniques. You may be surprised to find that some ideas are even workable. In any case, your clients should feel that you share their enthusiasm and excitement about their kitchen, breakfront or whatever, and that you are proud to be a party to its creation. Remind yourself that no matter how many kitchens or other projects you have done, this particular one will be unique and dear to the people who have trusted you to build it. After all, if they didn't feel so strongly about their furnishings, they wouldn't have come to you, the master custom cabinetmaker, in the first place.

Having received a deposit and begun working on the project, stay in close touch with the anxious owners-to-be. At appropriate stages, invite them into the shop to view the progress. The idea is to allow them to share the excitement of creation; you also want them to remember to pay the agreed-upon installments without prompting.

When the project is finally complete and awaiting delivery, call your clients in for a viewing. Before they arrive, cover the work with a clean white sheet. This conveys to them that you are protecting the piece from the perils of the workplace and also adds to the drama of the unveiling. This viewing allows everybody to discuss and enjoy the completion of the project free of the usual hustle and tension of installation. Once again, you are doing what you can to promote good feelings.

When installation time arrives, observe the basic rules of etiquette. Avoid trashing your clients' doors, floors, carpets or poodles. (I have done in one of each over the past 20 years.) Protect vulnerable areas with drop cloths or moving blankets (the latter can be rented or bought from local U-Haul outlets). Give a bone to the poodle and lock it in the bathroom. Before leaving, vacuum and wipe clean everything you touched and put everything you disturbed back in its place. The only evidence of your presence should be the professionally installed project. If the project was a kitchen, leave a fresh bouquet of flowers on the sparkling new countertop. Leave several business cards tucked in the silver drawer, and remember to let the mutt out of the bathroom.

Follow-Up

Summoning the energy to follow up can sometimes be hard to do, but it's definitely worthwhile. A month or so after the project has been completed, when your clients have had a chance to live with the work, give them a call and ask how everything is going. Chances are excellent that everything will be just fine. (If it wasn't, you assuredly would have heard by now.) This follow-up phone call is much appreciated and will reinforce your clients' perception that you care about the quality of your work and their satisfaction with it beyond the final payment. It is a superb way to turn a former client into a future client.

Working for the Generals

In new homes and large remodels, you may find yourself working for a general contractor. Maintaining a good reputation with this particular breed of client requires a different form of etiquette. Leaving flowers on the countertop just will not work.

Establishing a relationship with a contractor is less a matter of achieving personal rapport then of striving to be a well-greased cog in the machine that built Jack's house. And that machine is, of course, a merry-go-round. On it you will find plumbers, electricians, carpenters, painters, wallpaper hangers, stairbuilders, tile setters, drywall installers, carpet-layers, architects, lost-looking owners and six different radios set to six different stations. Your job here is to become a good carousel horse, running like hell in one place and never getting in anyone's way. To this end, consider the golden rule of subcontracting: Take your orders from the general, but do nothing until you've talked to the other subs first. These are the people you will actually have to work around (and over, under and beside). Be assured that if you work well with the other subcontractors on the project, you'll look just fine to the top gun. His or her main concern is that the job as a whole progress smoothly, so don't stop the music (in fact, bring your own radio).

It also pays to keep the army of subcontractors happy with you as well, since you will probably find yourself working around them again. Remember that all subcontractors are essentially in the same business, and that work for one tends to lead to work for everyone else. I always look forward to installation time and the sharing of bad coffee and even worse construction-site humor with other tradesmen whose respect I have earned.

Working with the Team

It may be clear on your contract who is paying your bill, but this is not necessarily the person who will be making critical decisions about the job. On larger projects, decisions concerning your piece of the pie might be made by any number of people, including the general contractor, architect, interior designer, finish subcontractor and owners. At any given time, who should make a decision depends largely upon the aspect of the project that is under question. The owner (if there is more than one, make sure they speak with one voice), in conjunction with the designer, should have the final say on matters concerning layout, design details and options that were called for in the final approved drawings for the project. If you were not the designer, be sure you are clear who was. If a correction affects another subcontractor, inform the general contractor immediately.

In reality, almost any change that occurs after the final drawings are approved will likely affect everyone. It is imperative that those involved in the direction of a large project meet together on a regular basis to keep the work on course. This is the only way you can be assured that the change orders you may receive (you already know they have to be in writing) are understood and approved by the people who are affected by them. If you suspect that the orders haven't been seen by everybody (I've been on a number of jobs undermined by interpersonal hostilities), go up the chain of command and ask. It will do your reputation no harm to question orders properly, but it can do you a world of harm if you are caught in the middle and blamed for a change that should not have been made.

Epilogue

Our work as independent custom cabinetmakers is a relatively rare phenomenon in a world that rushes headlong toward the 21st century. As we become a nation of consumers and service people for the rest of the planet, the number of people who derive a livelihood from the creative use of their minds and hands will likely continue to diminish. Our "working at woodworking" will never be found on the charts that list growth industries. Yet the ultimate human value of participating in an occupation that walks tenuously between the handicrafts of the last century and the electronic-robotic revolution of the next can hardly be judged by its impact on the gross national product.

This is important work that we are doing. The lifestyle that comes to us through the nature of this work nurtures and enriches us. And I trust that the products that we build and offer to our neighbors will enrich their lives as well.

Appendix I: Tools and Supplies

Many of the products listed here may be found elsewhere, but I have had good results when dealing with the following companies.

Back-to-back clamps
Manufactured by Griset Industries, P.O. Box 10114, Santa Ana, CA 92711; available through Woodworker's Supply and Trendlines (see addresses under "Tools").

Cabinet jacks
Gillift, 1605 N. River, Independence, MO 64050.

Contact cement, water-based
3M. Call (800) 666-6477 for local distributor.

Drill guides for 32mm-system holes
Double-drilling bar unit manufactured by J & R Enterprises, 12629 N. Tatum Blvd., #431, Phoenix, AZ 85032. Call (602) 953-0178.

Single-bar units manufactured by Häfele Co., 3091 Cheyenne Dr., Archdale, NC 27263; available direct from Häfele or through mail-order tool suppliers.

Face-frame boring bit
Manufactured by Ritter, Inc., 521 Wilbur Ave., Antioch, CA 94509; available through Woodworker's Supply (see address under "Tools").

Finishing materials, water-based
Hydrocote Co., Inc., P.O. Box 160, Tennent, NJ 07763. Call (800) 229-4937 for local distributor.

Glue applicators
Self-wetting roller: Woodworker's Supply (see address under "Tools").

Dual-vented tip (for slots) made by Lamello; available from Highland Hardware (see address under "Tools").

Hinge-boring and insertion machine, line bore
Blum Junior minipress manufactured by Julius Blum, Inc. Call (800) 438-6788 for local distributor.

Ritter Mini Drill manufactured by Ritter, Inc., 521 Wilbur Ave., Antioch, CA 94509.

Jigs
Jig to locate shank holes for door and drawer pulls manufactured by Häfele Co.; available direct from Häfele (see address under "Drill guides") or through Woodworker's Supply (see address under "Tools").

Hinge-mounting plate jig, drawer-slide jig for face-frame applications and centering pins for locating shank holes for drawer-face adjustable cams all manufactured by Blum. Call (800) 438-6788 for local distributor.

Pneumatic nail guns
Senco. Call (800) 543-4596 for local distributor.

Radial-arm-saw sliding stop
Available from Biesemeyer, 216 S. Alma School Rd., Mesa, AZ 85202. Includes right- to-left-reading stick-on tape measure for fence.

Router accessories
Grooving bit with pilot bearing on shank manufactured by Ocemco. Call (800) 237-8613 for local distributor.

Kerfing cutter, to produce 7/32-in. groove for undersized 1/4-in. panels, manufactured by Ocemco (see above).

Drawer-lock bit manufactured by Freud. Available from Woodworker's Supply and Trendlines (see addresses under "Tools").

Router plate available from Excalibur (see address under "Table-saw rip fences").

Safety gear
Work apron, respirator and safety glasses: Bridge City Tool Works, 1104 N.E. 28th Ave., Portland, OR 97232.

Earplugs: My favorite are made by Moldex-Metric Inc., available at local lumberyards.

Dust mask: Double-strap, double-thickness #8560 by 3M. Call (800) 666-6477 for local distributor.

First aid: American Red Cross. Call (800) 322-8349 for information on local chapters and availability of books and courses.

Sawblades
Best blade for all-around use: Freud No. LU98 M010, or Forrest Duraline HI-AT.

Sharpening accessories
Available from most mail-order sources (see "Tools").

Buffing compounds available from Rio Grande Supply, 6901 Washington N. E., Albuquerque, NM 87109, and other jeweler-supply houses.

Shop-helper hold-downs
Manufactured by Western Commercial Products, P.O. Box 238, Tulare, CA 93275; available through Woodworker's Supply and Trendlines (see addresses under "Tools").

Stikit sandpaper and adapting pads for electric sanders
Manufactured by 3M; available through its distributors, also through Trendlines (see address under "Tools").

System hardware and fastenings
Drawer slides, hinges, connector screws and fittings, cover caps, drawer-face-adjustment cams, adjustable cabinet feet, door and drawer bumpers all available through Julius Blum, Inc. Call (800) 438-6788 for catalog and local distributor. Also from Häfele, (800) 334-1873.

Confirmat knockdown fittings manufactured by Häfele (see address under "Drill guides").

Table-saw rip fences
Biesemeyer
216 S. Alma School Rd.
Mesa, AZ 85202

Delta
246 Alpha Dr.
Pittsburgh, PA 15238

Excalibur
3241 Kennedy Rd.
Scarborough, Ontario
Canada M1V 259

Quintec Paralok
5728 N.E. 42nd Ave.
Portland, OR 97218

Vega Enterprises
Rt. 3, Box 193
Decatur, IL 62526

Tools
Highland Hardware
1045 N. Highland Ave.
Atlanta, GA 30306

Robert Kaune (antique tools)
511 W. 11th St.
Port Angeles, WA 98362

Tool Crib
Box 1716
Grand Forks, ND 58206

Trendlines
375 Beacham St.
Chelsea, MA 02150

Whole Earth Access
2990 7th St.
Berkeley, CA 94710

Woodcraft Supply
P.O. Box 1686
Parkersburg, WV 26101

Woodworker's Supply of New Mexico
5604 Alameda Pl. N.E.
Albuquerque, NM 87113

Vix bits
Available through most mail-order tool suppliers (see "Tools").

Water level, reservoir type
Manufactured and available direct from Price Brothers Tools,
1053 Sixth St., Novato, CA 94945.

Appendix II: Further Reading

Business
Clifford, Denis and Ralph Warner. *The Partnership Book.* Berkeley, CA: Nolo Press, 1987.

Kamoroff, Bernard. *Small-Time Operation: How to Start Your Own Small Business, Keep Your Books, Pay Your Taxes and Stay out of Trouble.* Laytonville, CA: Bell Springs Publishing, 1988.

Philips, Michael and Salli Rasberry. *Marketing Without Advertising.* Berkeley, CA: Nolo Press, 1986.

Warner, Ralph. *Everybody's Guide to Small Claims Court.* Berkeley, CA: Nolo Press, 1989.

Publications of the Small Business Administration. Call (800) 368-5855 for information.

Federal tax information and free pamphlets. Call (800) 424-1040.

General information. National Association for the Self-Employed, 2324 Gravel Road, Fort Worth, TX 76118. Call (800) 232-6273

Design
Levine, Paul. *Making Kitchen Cabinets.* Newtown, CT: The Taunton Press, 1988.

Stem, Seth. *Designing Furniture.* Newtown, CT: The Taunton Press, 1989.

Selecting a dust collector
Fine Woodworking #67, pp. 70-75.

Selecting a sawblade
Fine Woodworking #70, pp. 36-41.

Selecting a surface planer
Fine Woodworking #52, pp. 72-78.
Fine Woodworking #65, pp. 102-104.

Selecting a table-saw rip fence
Fine Woodworking #68, pp. 41-45.

Tuning up and selecting bandsaws
Fine Woodworking #63, pp. 62-67.
Fine Woodworking #75, pp. 75-77.

Tuning up and selecting radial-arm saws
Fine Woodworking #73, p. 67.
Fine Woodworking #78, pp. 60-66.

Tuning up and selecting table saws
Fine Woodworking #56, pp. 50-57.
Fine Woodworking #70, p. 37.
Fine Woodworking #78, p. 69.

Appendix III: Production Flow Chart

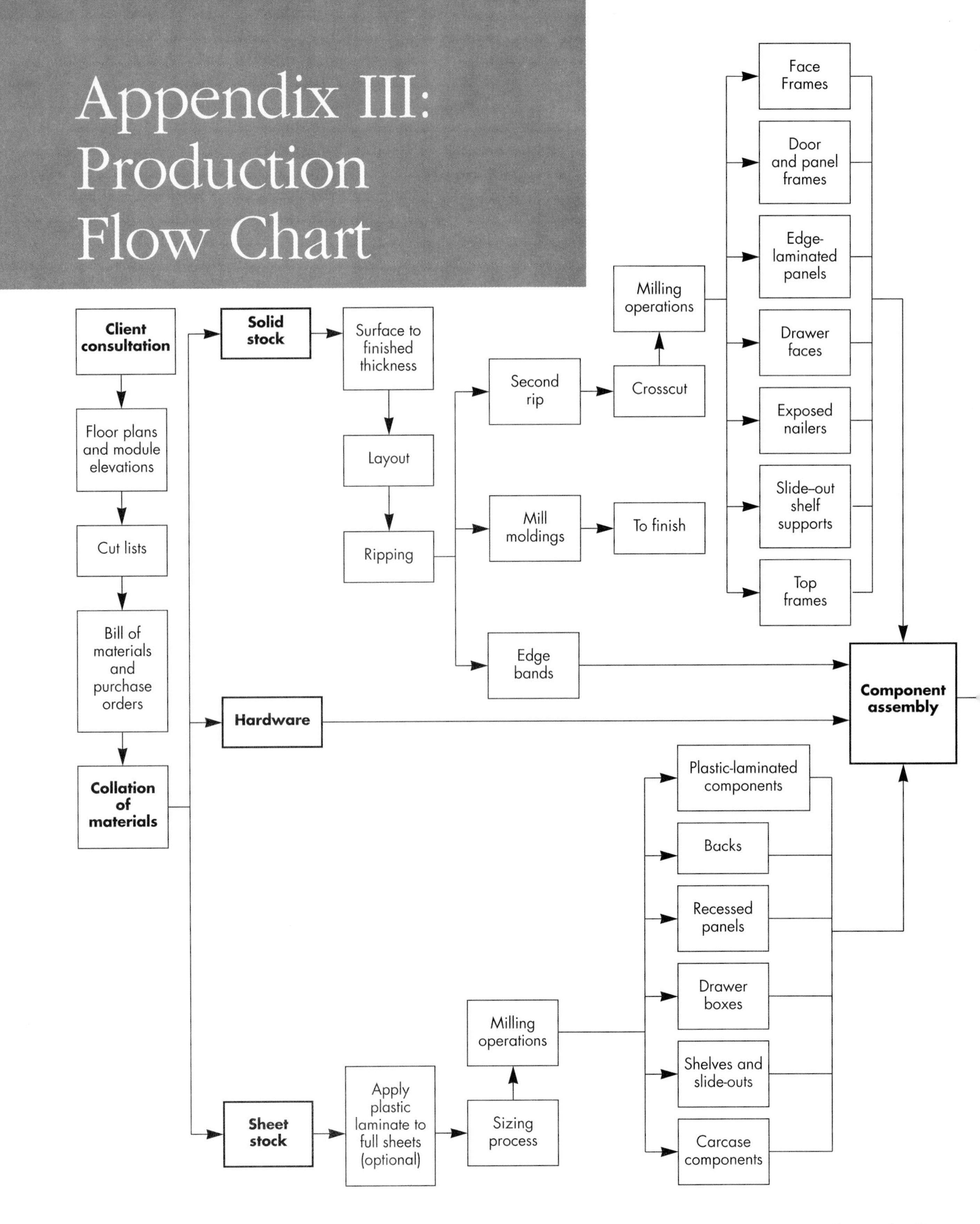

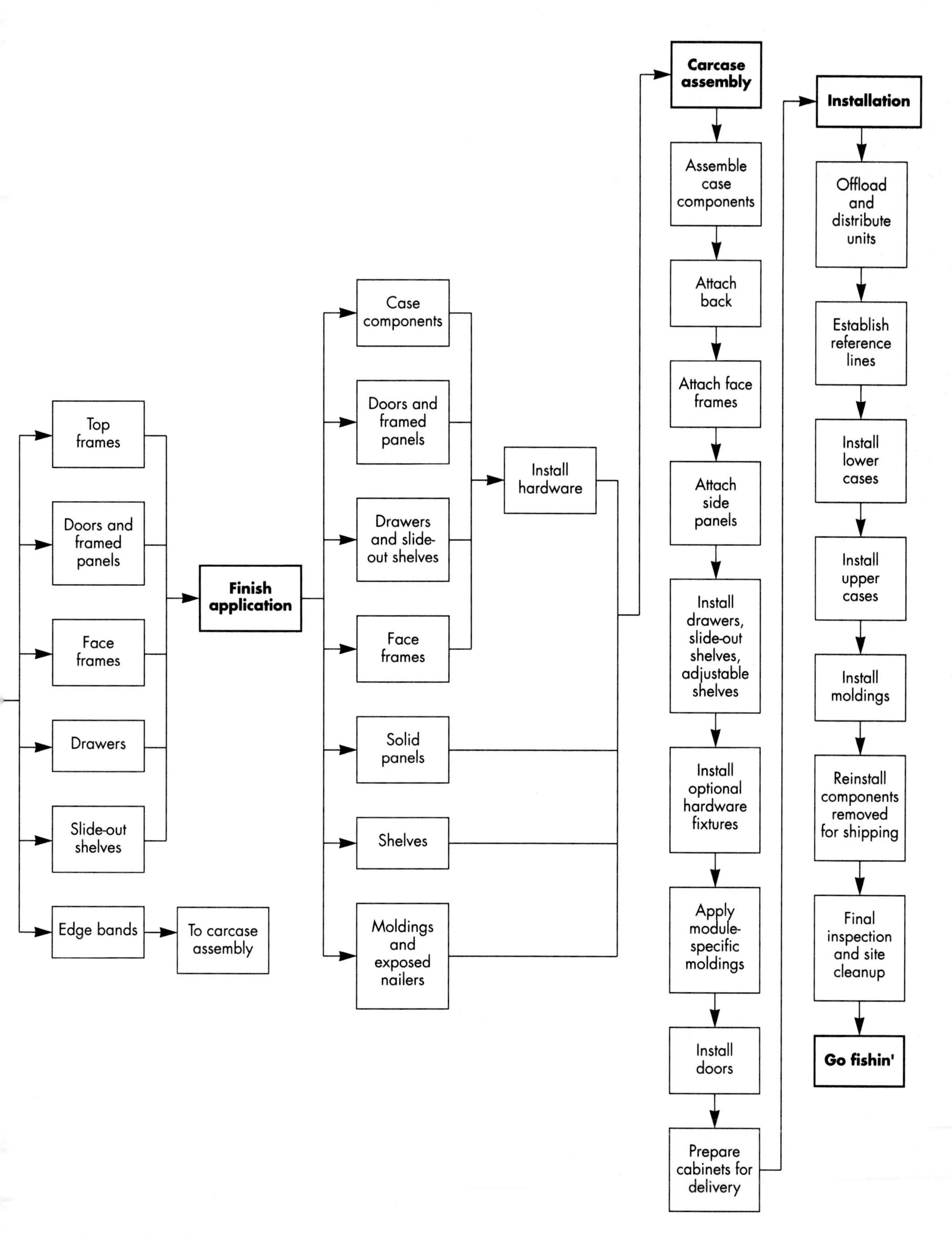

Top frames
Doors and framed panels
Face frames
Drawers
Slide-out shelves
Edge bands
To carcase assembly
Finish application
Case components
Doors and framed panels
Drawers and slide-out shelves
Face frames
Solid panels
Shelves
Moldings and exposed nailers
Install hardware
Carcase assembly
Assemble case components
Attach back
Attach face frames
Attach side panels
Install drawers, slide-out shelves, adjustable shelves
Install optional hardware fixtures
Apply module-specific moldings
Install doors
Prepare cabinets for delivery
Installation
Offload and distribute units
Establish reference lines
Install lower cases
Install upper cases
Install moldings
Reinstall components removed for shipping
Final inspection and site cleanup
Go fishin'

Index

Editor: Laura Tringali
Designer: Deborah Fillion
Layout artist: Cathy Cassidy
Illustrator: Vince Babak
Photographer: Patrick Cudahy
Copy/production editor: Peter Chapman

Typeface: Garamond
Paper: Warrenflo, 70 lb., neutral pH
Printer and binder: Arcata Graphics

B2AM

Fine WoodWorking

1 year (6 issues) for just $25.00 — over 24% off the newsstand price

Outside the U.S. $30/year (U.S. funds, please) Canadian residents add 7% GST.

Use this card to subscribe, or request additional information about Taunton Press books and videos.

☐ Payment enclosed
☐ Bill me
☐ Please send me your *Fine Woodworking* book and video catalog.

Name ______________________

Address ______________________

City ______________________

State ______________________

Zip ______________________

Phone ______________________

I am a ☐ novice ☐ intermediate ☐ advanced ☐ professional woodworker.

B2AM

Fine WoodWorking

1 year (6 issues) for just $25.00 — over 24% off the newsstand price

Outside the U.S. $30/year (U.S. funds, please) Canadian residents add 7% GST.

Use this card to subscribe, or request additional information about Taunton Press books and videos.

☐ Payment enclosed
☐ Bill me
☐ Please send me your *Fine Woodworking* book and video catalog.

Name ______________________

Address ______________________

City ______________________

State ______________________

Zip ______________________

Phone ______________________

I am a ☐ novice ☐ intermediate ☐ advanced ☐ professional woodworker.

NO POSTAGE
NECESSARY
IF MAILED
IN THE
UNITED STATES

BUSINESS REPLY MAIL
FIRST CLASS MAIL PERMIT No. 19 NEWTOWN, CT
POSTAGE WILL BE PAID BY ADDRESSEE

The Taunton Press
63 South Main Street
P.O. Box 5506
Newtown CT 06470-9971

TAUNTON
MAGAZINES
...by fellow enthusiasts

NO POSTAGE
NECESSARY
IF MAILED
IN THE
UNITED STATES

BUSINESS REPLY MAIL
FIRST CLASS MAIL PERMIT No. 19 NEWTOWN, CT
POSTAGE WILL BE PAID BY ADDRESSEE

The Taunton Press
63 South Main Street
P.O. Box 5506
Newtown CT 06470-9971

TAUNTON
MAGAZINES
...by fellow enthusiasts

NO POSTAGE
NECESSARY
IF MAILED
IN THE
UNITED STATES

BUSINESS REPLY MAIL
FIRST CLASS MAIL PERMIT No. 19 NEWTOWN, CT
POSTAGE WILL BE PAID BY ADDRESSEE

The Taunton Press
63 South Main Street
P.O. Box 5506
Newtown CT 06470-9971